STUDENT'S STUDY GUIDE AND SOLUTIONS MANUAL

to accompany

USING AND UNDERSTANDING MATHEMATICS

A QUANTITATIVE REASONING APPROACH

STUDENT'S STUDY GUIDE AND SOLUTIONS MANUAL

to accompany

USING AND UNDERSTANDING MATHEMATICS

A QUANTITATIVE REASONING APPROACH

Jeffrey O. Bennett
University of Colorado at Boulder

William L. Briggs
University of Colorado at Denver

ADDISON-WESLEY

An imprint of Addison Wesley Longman, Inc.

Reading, Massachusetts • Menlo Park, California • New York • Harlow, England
Don Mills, Ontario • Sydney • Mexico City • Madrid • Amsterdam

ISBN 0-201-59084-0

2 3 4 5 6 7 8 9 10 VG 0099

Acknowledgments

I am grateful to W.E. Briggs, Steve Ouellette, and Paul Lorczak for checking and proofreading the solutions in this book. Thanks also to Wesley Johnson for proofreading the text of the manual. George Nichols deserve high praise for drawing the figures for the solutions. Elka Block, Kim Ellwood, Rachel Reeve, and Joe Vetere at Addison Wesley Longman provided valuable assistance in producing the book. I also thank Julie and Katie whose patience and understanding allowed this manual to be completed on a very short time scale.

Table of Contents

STUDENT'S STUDY GUIDE AND SOLUTIONS MANUAL

to accompany

USING AND UNDERSTANDING MATHEMATICS

A QUANTITATIVE REASONING APPROACH

INTRODUCTION

Welcome to the *Study Guide and Solutions Manual* for *Using and Understanding Mathematics*. Hopefully, with the help of this book, your course in quantitative reasoning will be both enjoyable and successful. The goal of this guide is not to add to your work load for the course, but to give you a set of concise notes that will make your studying as effective as possible. If you work with this guide as you read and do problems, you should get the most from the course.

This guide is organized according to the units in the textbook. For each unit you will find the following features:

Overview

This section provides a brief survey of the unit, its major points and its goals. This feature is not a substitute for reading the text!

Key Words and Phrases

This section is simply a list of the important words and phrases used in the unit. You can use this section for review, study, and self-testing. You should be able to explain or define all of the terms on this list.

Key Concepts and Skills

In this section you will find a summary of the most important concepts in each unit. These concepts may be general ideas (for example, the distinction between deductive and inductive arguments) or basic skills (for example, creating the equation of a straight line). This section should also be helpful for review, study, and self-testing.

Solutions to (Most) Odd Problems

Each unit of this guide ends with the solutions to most of the odd-numbered problems in the text book. Mathematics is not a spectator sport! Reading the solutions is never a substitute for working the problems. *You are strongly advised to work the problems first and then check the solutions.*

How to Succeed In This Course

Using *This* Book

Before we get into more general strategies for studying, here are a few guidelines that will help you use *this* book most effectively.

- Before doing any assigned problems, read assigned material *twice*:
  On the first pass, read quickly to gain a "feel" for the material and concepts presented.
  On the second pass, read the material in more depth, and work through the examples carefully.
- During the second reading, take notes that will help you when you go back to study later. In particular:
  Use the margins! The wide margins in this textbook are designed to give you plenty of room for making notes as you study.
  Don't highlight — underline! Using a pen or pencil to underline material requires greater care than highlighting, and therefore helps to keep you alert as you study.
- After you complete the reading, and again when studying for exams, make sure you can answer the *review questions* at the end of each unit.
- You'll learn best by *doing*, so work plenty of the end-of-unit problems. Don't be reluctant to work more than the problems that your instructor assigns.

Budgeting Your Time

A general rule of thumb for college classes is that you should expect to study about 2 to 3 hours per week *outside* class for each unit of credit. For example, a student taking 15 credit hours should spend 30 to 45 hours each week studying outside of class. Combined with time in class, this works out to a total of 45 to 60 hours per week — not much more than the time required of a typical job. Moreover, except for class time, you get to choose your own hours. Of course, if you are working while you attend school, you will need to budget your time carefully.

If you find that you are spending fewer hours than these guidelines suggest, you can probably improve your grade by studying more. If you are spending more hours than these guidelines suggest, you may be studying inefficiently; in that case, you should talk to your instructor about how to study more effectively for a mathematics class.

General Strategies for Studying

- Don't miss class. Listening to lectures and participating in discussions is much more effective than reading someone else's notes. Active participation will help you retain what you are learning.
- Budget your time effectively. An hour or two each day is more effective, and far less painful, than studying all night before homework is due or before exams.

- If a concept gives you trouble, do additional reading or problem solving beyond what has been assigned. If you still have trouble, *ask for help*: you surely can find friends, colleagues, or teachers who will be glad to help you learn. Never be reluctant to ask questions or ask for help in this course. If you have a question or problem, it is extremely unlikely that you will be alone!
- Working together with friends can be valuable; you improve your own understanding when discussing concepts with others. However, be sure that you learn *with* your friends and do not become dependent on them.

Preparing for Exams

- Rework problems and other assignments; try additional problems to be sure you understand the concepts. Study your performance on assignments, quizzes, or exams from earlier in the semester.
- Study your notes from lectures and discussions. Pay attention to what your instructor expects you to know for an exam.
- Reread the relevant sections in the textbook, paying special attention to notes you have made in the margins.
- Study individually before joining a study group with friends. Study groups are effective only if *every* individual comes prepared to contribute.
- Don't stay up too late before an exam. Don't eat a big meal within an hour of the exam (thinking is more difficult when blood is being diverted to the digestive system).
- Try to relax before and during the exam. If you have studied effectively, you are capable of doing well. Staying relaxed will help you think clearly.

Finally, good luck! We wish you a memorable and beneficial experience in quantitative reasoning.

1. PRINCIPLES OF REASONING

Overview

Before discussing Chapter 1, we urge you to take a few minutes to read the prologue to the text book. This short chapter sets the stage for the entire book. It presents the idea of quantitative reasoning and discusses the importance of interdisciplinary thinking. It gives a high altitude picture of mathematics and how it impacts many other subjects that you will encounter either in other courses or in your career. Finally, it gives some advice on using the book and studying for your course. It's worth a quick reading. Now on to Chapter 1.

We have taught this course to many students of over many years. Often the most serious weakness that students bring to the course is not poor mathematical skills, but poor reasoning skills. Often it's not multiplying two numbers that creates problems, but deciding which two numbers to multiply! For this reason, the book opens with a chapter that contains virtually no mathematics. The emphasis of the chapter is critical thinking and logical skills.

In this chapter you will encounter some introductory logic, but don't worry, we don't get carried away with symbolic logic and heavy-duty truth tables. In fact, much of this chapter may be familiar to you from previous courses in logic or philosophy.

Unit 1A The Forces of Persuasion

Unit 1A (*The Forces of Persuasion*) opens the chapter by exploring common **fallacies** that you might encounter in advertizing or (bad) news reports. We present a dozen different so-called informal fallacies, some of which may seem quite obvious, others of which may be quite subtle. Critical reading and thinking will help you avoid becoming a victim of these fallacies!

Key Words and Phrases

appeal to popularity	false cause	hasty generalization
appeal to ignorance	limited choice	appeal to emotion
appeal to force	personal attack	circular reasoning
subjectivism	diversion	straw man

Key Concepts and Skills
- identify informal fallacies in advertisements and news reports.

Unit 1A Solutions

Identifying Fallacies.

13. Without further historical substantiation, it is a <u>false cause fallacy</u> to claim that Reagan was responsible for the demise of the Soviet Union.

15. This question is a form of <u>circular reasoning</u> (begging the question) called a <u>complex question</u>. To someone listening to the question (for example, a jury) the conclusion is almost inevitable.

17. This statement is a circumstantial case of a <u>personal attack.</u> The fact that another person smokes is irrelevant to one's own personal decision to smoke.

19. This is an example of <u>appeal to ignorance</u>: an absence of proof for one conclusion (telepathy does not exist) does not prove the opposite conclusion (telepathy does exist).

21. This is a case of <u>personal attack</u>: the fact that the Senator gets campaign funds from the NRA does not necessarily mean that he or she will only introduce bills that are to the NRA's liking.

23. This is an example of <u>false cause</u>: the mere fact that there are now more overweight people and fewer smokers does not prove that quitting smoking causes overeating.

Unit 1B Propositions – Building Blocks of Logic

Overview

In this unit we introduce formal logic in a very casual way. We start with **propositions** — statements that make a claim that can be true or false. Then we look at the **connectors** that can be used with propositions to make more complex propositions. The connectors that you will encounter are

- *not* (**negation**)
- *or* (**disjunction**)
- *and* (**conjunction**)
- *if ... then* (**implications**).

Whereas many logic books, make heavy use of symbolic logic and truth tables, we will use truth tables primarily for fairly simple propositions that involve one, two, or three connectors. So our excursion into symbolic logic will be limited and designed to provide only an introductory glimpse.

The *if ... then* connector is quite important in both logic and everyday speech (for example, *if I pass this course, then I will graduate*). For this reason, we spend a little time discussing other forms of the proposition *if P, then Q*. These other forms are called the

- **converse** (*if Q, then P*)
- **inverse** (*if not P, then not Q*)
- **contrapositive** (*if not Q, then not P*).

This particular discussion may seem a bit technical, but it's also extremely practical. For example, suppose it's true that *if I read the book, then I will pass the course*. Does it follow that *if I don't read the book, then I won't pass the course*? You will see!

The last step is to explore some special forms of propositions that the Greeks introduced over 2000 years ago. There are four of these basic **categorical propositions**:

- All S are P (for example, *all barbers are lawyers*)
- No S are P (for example, *no birds are movie stars*)
- Some S are P (for example, *some monkeys are vegetarians*)
- Some S are not P (for example, *some fish are not doctors*).

While these four forms may seem rather specialized, it turns out that they can be used to express many everyday statements. Although propositions involve words, you will learn how to draw pictures (**Venn diagrams**) to work with propositions.

Key Words and Phrases

proposition	negation	conjunction
disjunction	conditional	antecedent
consequent	converse	inverse
contrapositive	categorical proposition	

Key Concepts and Skills

- negation, conjunction, disjunction, and conditionals from propositions and determine truth values.
- use of truth tables to evaluate the truth of compound propositions that use two or more connectors.
- analysis of various forms of *if .., then* propositions.
- Venn diagrams for four categorical propositions.

Unit 1B Solutions

Identifying Propositions.

1. This statement is a proposition. It has a subject and a predicate, and makes a claim that is capable of being true or false.

3. This statement is a proposition; in fact, it is a categorical proposition.

5. This imperative statement is not a proposition.

Negating Propositions.

7. The negation is *The Gettysburg Address was not given by George Washington*. The original proposition is false, so its negation is true.

9. The negation is *Mark Twain did not write Tom Sawyer*. The original proposition is true, so its negation is false.

11. The negation is *New York is the capital of the United States*. The original proposition is true, so its negation is false.

13. The negation is *some snakes are not mammals*. The original proposition is false, so its negation is true.

Multiple Negatives.

15. The President's veto of the bill is in opposition to the bill. Therefore an override of the veto supports the bill.

17. A repeal of the affirmative action laws would oppose affirmative action. The Dean's opposition of the repeal means the Dean supports affirmative action.

19. There are two negations: opposes and ban. If the Senator supported the ban on anti-war demonstrations, he would oppose demonstrations. Because he opposes the ban, he supports anti-war demonstrations (which says nothing about his stand on the war issue itself).

Truth of "And" Statements

21. Both propositions are true, so the compound statement is true.

23. Both statements cannot be true, so the compound statement is false.

Interpreting "Or"

25. The statement *the menu offers a choice of appetizer or dessert* uses *or* in the exclusive sense. It is generally understood that patrons can choose an appetizer or a dessert, but not both.

27. The statement *if I win the lottery, I will go to Brazil or Nepal* may be interpreted using the rule of thumb that the *or* is inclusive unless understood otherwise. Thus, you will be going to either Brazil or Nepal or both. It may also be interpreted in the exclusive sense meaning that you will be going to only one or the other of Brazil or Nepal.

29. The statement *the road will be made of asphalt or concrete* needs to be further qualified before determining if the *or* is meant to be inclusive or exclusive. The road could be made of both asphalt and concrete, or just one or the other.

Truth of "Or" Statements

31. Both propositions are true, so the compound statement is true.

33. Both statements are false, so the compound statement is false.

35. The first statement is true, the second statement is false, so the compound statement is true.

Truth of Compound Statements

37. Given that p is true and q is false, $\sim q$ is true, and p and $\sim q$ is true.

39. Given that p is true and q is false, $\sim q$ is true, and p or $\sim q$ is true.

41. Given that p is true, q is false, and r is true, p and q is false, and $(p$ and $q)$ and r is also false.

Truth Tables for Compound Propositions

43.

p	q	$\sim q$	p or $\sim q$
T	T	F	T
T	F	T	T
F	T	F	F
F	F	T	T

The statement is true in all cases except when p is false and q is true.

45.

p	q	$\sim p$	$\sim p$ or q
T	T	F	T
T	F	F	F
F	T	T	T
F	F	T	T

The statement is true is all cases except when p is true and q is false.

47.

p	q	r	p or q	$(p$ or $q)$ and r
T	T	T	T	T
T	T	F	T	F
T	F	T	T	T
T	F	F	T	F
F	T	T	T	T
F	T	F	T	F
F	F	T	F	F
F	F	F	F	F

49.

p	q	r	~r	p or q	(p or q) and ~r
T	T	T	F	T	F
T	T	F	T	T	T
T	F	T	F	T	F
T	F	F	T	T	T
F	T	T	F	T	F
F	T	F	T	T	T
F	F	T	F	F	F
F	F	F	T	F	F

51. Interpreting Connectors.

a. The number of people who eat WonderCorn is the sum of the people who eat WonderCorn only and the people who eat both. Thus, 24 + 18 = 42 eat WonderCorn.

c. The number of people who eat WonderCorn *and* PrimePop is given as 18.

Truth of "If..Then" Statements

53. The antecedent is *London is the capital of England*; the consequent is *Paris is the capital of China*. Only the antecedent is true. The entire statement is false because when the antecedent is true and the consequent is false, the whole implication is false

55. The antecedent is *Boston is the capital of Colorado*; the consequent is *Moscow is the capital of France*, both of which are false. Because the implication relies on a false condition (*Boston is the capital of Colorado*), anything can follow and the whole implication remains true. The entire statement is true.

57. The antecedent is $8 \times 4 = 32$; the consequent is $8 \times 8 = 64$. The antecedent and the consequent are both true; therefore the entire statement is true.

Symbolic "If...Then" Statements

59. Given that p is true and q is false, $\sim p$ is false, and $\sim p \Rightarrow q$ is true.

61. Given that p is true, q is false, and r is true, p or q is true, and $(p$ or $q) \Rightarrow r$ is true.

Truth Tables for "If...Then" Statements

63.

p	q	r	q and r	p ⇒ (q and r)
T	T	T	T	T
T	T	F	F	F
T	F	T	F	F
T	F	F	F	F
F	T	T	T	T
F	T	F	F	T
F	F	T	F	T
F	F	F	F	T

65.

p	q	r	p and q	(p and q) ⇒ r
T	T	T	T	T
T	T	F	T	F
T	F	T	F	T
T	F	F	F	T
F	T	T	F	T
F	T	F	F	T
F	F	T	F	T
F	F	F	F	T

Converse, Inverse, and Contrapositive

67. The converse is *if Marco lives in the United States, then he lives in Chicago*. The inverse is *if Marco does not live in Chicago, then he does not live in the United States*. The contrapositive is *if Marco does not live in the United States, then he does not live in Chicago*.

69. The converse is *if I got wet, then I went swimming*. The inverse is *if I do not go swimming, then I will not get wet*. The contrapositive is *if I do not get wet, then I did not go swimming*.

71. The converse is *if it's cold-blooded, then it's a reptile*. The inverse is *if it is not a reptile, then it is not cold-blooded*. The contrapositive is *if it is not cold-blooded, then it's not a reptile*.

Conditional Propositions.

73. This statement can be rephrased as *if a person is a member of Congress, then that person is a lawyer*. The antecedent is *a person is a member of Congress,* and the consequent is *that person is a lawyer*. There are cases in which the antecedent is true (a person is a member of Congress), but the consequent is false (that person is not a lawyer). This combination of truth values makes the conditional statement false.

75. This statement can be rephrased as *if a person is a musician, then that person can play the saxophone*. The antecedent is *a person is a musician,* and the consequent is *that person can play the saxophone*. There are certainly some musicians that cannot play the saxophone. So the antecedent can be true, while the consequent is false. This makes the conditional statement false.

Working with Categorical Propositions.

77. (i) no rephrasing necessary. (ii) The subject is *police officers* and the predicate is *women* .(iii) The Venn diagram is shown below. We know only that there are police officers in the overlapping region of the circles.

79. (i) The proposition might be rephrased slightly as *no*

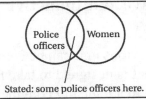

Stated: some police officers here.

Republicans are socialists. (ii) The subject is *Republican* and the predicate is *socialist* . (iii) The Venn diagram is shown below.

81. (i) The proposition *no bachelors are married* can be

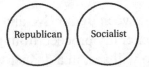

rephrased as *no bachelors are married people*. (ii) The subject is *bachelors* and the predicate is *married people*. (iii) The Venn diagram is shown below.

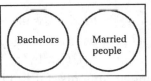

83. (i) The proposition is in standard form if we realize that Ronald Reagan in a singular set. (ii) The subject is *Ronald Reagan* and the predicate is *great presidents*. (iii) The Venn diagram is shown below.

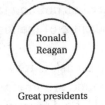

Great presidents

85. Another Use of Venn Diagrams.

a. The total number of people in the room is $7 + 6 + 9 + 6 = 28$.

b. There are $6 + 9 = 15$ women, 6 of whom are Republicans, and 9 of whom are not Republicans.

c. All numbers outside of the *women* circle correspond to men; so there are $7 + 6 = 13$ men in the room.

d. There are $7 + 6 = 13$ Republicans in the room, 6 of whom are women, 7 of whom are men.

e. There are 7 Republican men.

f. There are 6 non-Republican men (the 6 outside of both circles).

g. There are 6 women Republicans (the 6 inside of both circles).

h. There are 9 non-Republican women.

87. Interpreting a Survey. This problem illustrates another use of Venn diagrams. The Venn diagram below helps answer the questions.

a. The people who read the NYT consist of four

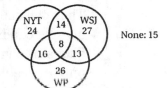

None: 15

categories: those who read the NYT only (24), those who read the NYT and the WSJ only (14), those who read the NYT and the WP only (16), and those who read all three (8). The total number of people who read the NYT is the sum of these four groups which equals 62.

c. The number of people who read all three papers is 8.

e. Those who read the NYT, but not the WP are those who read the NYT only (24) and those who read the NYT and WSJ only (14), for a total of 38 people.

Unit 1C Logical Arguments: Deductive and Inductive

Overview

The propositions that we studied in the previous unit can be combined in various ways to form **arguments**. An argument is designed to take the reader from several **premises** to a **conclusion** in a logically convincing way. In this unit we will be concerned with the analysis of arguments, a process that will determine whether an argument is reliable and should be believed. As you will see, even the most persuasive arguments can fail in many subtle ways.

Of primary importance is the distinction between **deductive** and **inductive** arguments. Deductive arguments generally proceed from general premises to a more specific conclusion. As we will see, in a deductive argument all of the premises are needed to reach the conclusion. By contrast, inductive arguments generally proceed from specific premises to a general conclusion.In an inductive argument the premises independently support the conclusion.

Much of this unit is spent analyzing three-line deductive arguments. They can consist of the four types of categorical propositions or they may involve conditional propositions (all studied in Unit 1B). Circle (or Venn) diagrams can be used to analyze these three-line arguments. Of fundamental importance in this business is the distinction between valid and invalid arguments. An argument is **valid** if, based on the circle diagram analysis, it is logically solid and consistent. An argument that fails the circle diagram analysis must contain a fallacy and is **invalid**. Validity has nothing to do with the truth of the premises or conclusion; it is a measure of the logical structure of the argument.

Having shown that an argument is valid, we can then ask if it is sound. A **sound** argument is valid and has true premises; a sound argument must lead to a true conclusion. Soundness is the highest test of a deductive argument.

Fallacies can arise in deductive arguments in many ways. Perhaps the most common fallacies occur in arguments that involve conditional (*if...then*) propositions. These fallacies appear in the everyday arguments of advertizing and news reports. There are four different forms of conditional arguments; two are valid and two are invalid:

- affirming the antecedent (valid)
- affirming the consequent (invalid)
- denying the antecedent (invalid)
- denying the consequent (valid).

Turning to inductive arguments, a most important observation is that validity and soundness don't apply. We ask about the **strength** or **weakness** of an inductive argument. Determining the strength of an inductive argument is often a subjective judgment, and so there are no systematic methods to apply. We close by illustrating how the interplay of inductive and deductive reasoning is important in mathematics.

Key Words and Phrases

premise	conclusion	deductive
inductive	valid	sound
strength	Pythagorean theorem	affirming the antecedent
affirming the consequent	denying the antecedent	denying the consequent

Key Concepts and Skills

- the distinction between deductive and inductive arguments.
- analysis of the validity of three-line deductive arguments consisting of categorical propositions, using circle diagrams.
- analysis of the validity of three-line deductive arguments involving conditional propositions, using circle diagrams.
- fallacies that arise in conditional arguments.
- the various combinations of valid/invalid and sound/unsound that can occur in deductive arguments.
- determining the soundness of three-line deductive arguments.
- determining the strength of inductive arguments.

Unit 1C Solutions

Deductive Arguments with Categorical Propositions.

51. (i)

Premise: All islands are tropical lands.

<u>Premise: All tropical lands are lands with jungles.</u>

Conclusion: All islands are lands with jungles.

(ii)

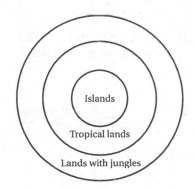

The argument is <u>valid</u> since according to the diagram all islands are lands with jungles.

(iii) Not all islands are tropical so the first premise is false. Also, some tropical lands (land between the tropics of Cancer (23.5° north latitude) and Capricorn (23.5° south latitude)), like the Sahara Desert are not lands with jungles, so the second premise is false. Even though the conclusion is true, the premises are false and the argument is <u>not sound</u>.

3. (i)

Premise: All dairy products are products containing protein.

<u>Premise: No soft drinks are products containing protein.</u>

Conclusion: No soft drinks are dairy products.

(ii)

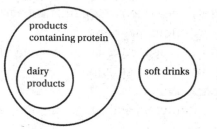

The conclusion follows from the diagram since the *soft drinks* circle does not overlap the *dairy products* circle. Thus the argument is <u>valid</u>.

(iii) The premises are both true; therefore the argument is <u>sound.</u>

5. (i)

Premise: No women are NFL quarterbacks.

<u>Premise: Some NFL quarterbacks are tall people.</u>

Conclusion: Some tall people are not women.

(ii)

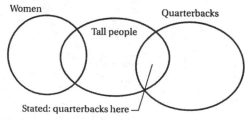

The conclusion is consistent with the diagram, so the argument is <u>valid</u>.

(iii) Both premises are true. Therefore the argument is <u>sound.</u>

7. (i)

Premise: Some lobbyists are people who work for the oil industry.

<u>Premise: All lobbyists are persuasive people.</u>

Conclusion: Some persuasive people are people who work for the oil industry.

(ii)

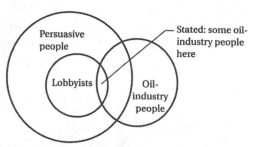

The diagram shows that there are oil industry people inside the *persuasive people* circle. Therefore, some persuasive people are oil industry people and the conclusion is consistent with the diagram. This means the argument is <u>valid</u>.

(iii) Since both premises are true and the argument is so the conclusion is not consistent with the diagram. Thus, the argument is <u>not valid</u>.

9. (i) We must first rephrase the premises so they fit the standard forms. The following argument is equivalent to the given argument.

Premise: No uninsured person is a person who can get medical treatment.

<u>Premise: Some people are uninsured people.</u>

Conclusion: Some people are not people who can get medical treatment.

(ii) There are three categories or groups in this argument: *uninsured people, people who can get medical treatment,* and *people.* The Venn diagram is shown below.

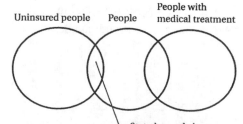

We see that the argument is <u>valid</u> since the diagram confirms the conclusion: there are *people* who are not in the *medical treatment* circle.

(iii) The first premise is false, since it is possible to get medical treatment without having insurance. Therefore, because of the false premise, the argument is <u>not sound</u>.

11. (i) Rephrasing is necessary for this argument:

Premise: All states in the EST zone are states east of the Mississippi River.

<u>Premise: Maine is a state in the EST zone..</u>

Conclusion: Maine is a state east of the Mississippi River.

(ii) The Venn diagram is shown below. We represent Maine by a circle even though it is a category with a single member. Notice that the Maine circle is placed inside of the inner circle representing *states in the EST zone*. This also places Maine inside of the circle for *states east of the Mississippi River.*

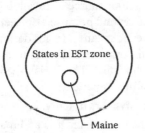

The conclusion of the argument is consistent with the diagram and the argument is <u>valid</u>.

(iii) The premises are both true, so this valid argument is also <u>sound</u>.

Valid and Sound.

13. A sound argument must be valid, so this combination is impossible.

15. Example of valid with two false premises and a true conclusion:

Premise: All mammals swim. (false)

<u>Premise: All fish are mammals. (false)</u>

Conclusion: All fish swim. (true)

Deductive Arguments with one Conditional.

17. (i) This argument does not need to be rephrased.

(ii) This is an example of affirming the antecedent.

(iii)

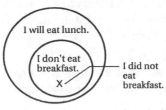

The first premise is diagrammed with the *I don't eat breakfast* circle inside of the *I will eat lunch* circle. This means that if you are inside of the *if I don't eat breakfast* circle (meaning you did not eat breakfast), then you are

necessarily inside of the *I will eat lunch* circle (meaning that you will eat lunch). The second premise refers to a single event (I did not eat breakfast) so it is diagrammed with an X inside of the *I don't eat breakfast* circle. The X is also inside of the *I will eat lunch* circle, so the conclusion follows. The argument is <u>valid</u>; it is an example of affirming the antecedent (or *modus ponens*).

19. (i) Rephrasing is not necessary for this argument.

(ii) This is an example of denying the antecedent.

(iii)

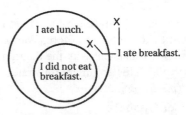

In this case, the X, representing the proposition *I ate breakfast,* can have two locations, both outside of the circle for *I did not eat breakfast*. The X inside the *I ate lunch* circle is not consistent with the conclusion which means the argument is <u>not valid</u>.

(iv) The fallacy in this argument is denying the antecedent which always leads to an invalid argument.

21. (i) A slight rephrasing helps for this problem:

Premise: If interest rates decline, then the bond market improves.

<u>Premise: Last week the bond market improved.</u>

Conclusion: Interest rates must have declined.

(ii) This is an example of affirming the consequent.

(iii)

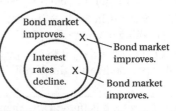

We see that of the two possible locations for the proposition *the bond market improved* (marked by the X), only one is consistent with the conclusion of the argument. Therefore the argument is <u>invalid</u> (the bond market could have improved for other reasons).

(iv) The fallacy in this argument is affirming the consequent.

23. (i) The argument can be rephrased:

Premise: If a person is a nurse, then s/he must know CPR.

Premise: Tom is a nurse.

Conclusion: Tom knows CPR.

(ii) This is an example of affirming the antecedent.

(iii)

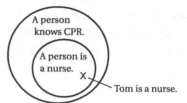

Being a nurse places Tom (the X) inside of both the *nurse* circle and the *CPR* circle. Therefore, Tom knows CPR, the conclusion is consistent with the diagram, and the argument is valid. This argument illustrates affirming the antecedent (*modus ponens*).

Chains of Conditionals.

29. (i) After rephrasing, the argument appears as a chain of conditional propositions:

Premise: If you shop, then I will make dinner.

Premise: If I make dinner, then you will take out the trash.

Conclusion: If you shop, then you will take out the trash.

(ii) The argument is valid since it has the valid form

Premise: If p, then q.

Premise: If q, then r.

Conclusion: If p, then r.

(iii) In order for the argument to be sound it must be valid and the premises must be true. The premises will be true if both people involved agree to the terms of the premises.

31. (i) The argument needs to be rephrased so that it appears as a chain on conditional propositions:

Premise: If a person is in the U.S., then that person has right to say anything at any time.

Premise: If a person has a right to say anything at any time, then that person has a right to yell "fire!" in a theater.

Conclusion: If a person is in the U.S., then that person has a right to yell "fire!" in a theater.

(ii) The argument is valid since it has the valid form

Premise: If p, then q.

Premise: If q, then r.

Conclusion: If p, then r.

(iii) In order for the argument to be sound it must be valid and the premises must be true. Most people would agree that the first premise is not true: a person does not have a right to say *anything* anytime. For example, slander is

considered to be at least unethical, if not illegal. Thus, the argument fails to be sound because one of its premises is false.

33. (i) Some rephrasing is helpful.

Premise: If taxes are cut, the U.S. government will have less revenue.

Premise: If the U.S. government has less revenue, then the deficit will be larger.

Conclusion: If taxes are cut, then the deficit will be larger.

(ii) It is valid because it has the form of a valid chain of conditionals.

(iii) The soundness of this argument depends on economic "truths." The first premise seems plausible since taxes are a major source of revenue for any government. The second premise also seems true since, in general, less revenue means a larger deficit (or a smaller surplus). Thus, the argument appears to be sound. On the other hand, economists and politicians cannot agree about the soundness of this argument!

Analyzing Inductive Arguments.

35. (i) Each of the premises is true and many more premises with the same property can be given in support of the conclusion. (ii) The many cases in which positive numbers a and b satisfy the property $(-a) \times (-b) > 0$ give this argument considerable strength. (iii) In fact, the conclusion of this argument is true. However, its truth must be established by proof and not with a long list of examples.

37. (i) The premise (which can be considered to be six separate premises) is true. (ii) However, it is easy to think of great composers whose names begin with letters other than B. Therefore, the argument is weak. (iii) The conclusion is clearly false.

Testing Mathematical Rules with Inductive Arguments.

39. Many different number pairs can be substituted into the expression $a + b = b + a$ and in all cases the equation will be satisfied. This amounts to a strong inductive argument for the truth of this statement.

41. One set of three numbers that do not satisfy the equation suffice to prove that the statement is false in general. For example $2^2 + 3^2 \neq (2 + 3)^2$.

Every Day Logic.

43. Because the argument generalizes from many specific instances, it is inductive. The conclusion is strengthened by each instance.

45. Because the argument generalizes from many specific instances, it is inductive. The conclusion is strengthened by each instance.

Unit 1D Analyzing Real Arguments

Overview

The arguments considered in this chapter so far have been rather unrealistic – after all, we don't read in news reports that *all fish can swim, all monkeys are fish, therefore all monkeys can swim*. These arguments are toy problems designed to develop and improve critical thinking skills. Having done our calisthenics on these simple arguments, it's now time to look at arguments of the real world; that is the goal of this unit.

Real world arguments are complicated by propositions that aren't in standard form, by **intermediate conclusions**, and by **assumed premises**. Many of the ideas from previous units cannot be applied directly, but they still have meaning. For example, it still helps to distinguish between deductive and inductive arguments, and the goal is still to determine if deductive arguments are sound and whether inductive arguments are strong.

One of the best techniques for analyzing such complex arguments is to draw a **flow chart** that shows how the premises are related and how they lead to the final conclusion. Even if you find it difficult to assess the soundness or strength of a complex argument, the process of drawing a flow chart can only improve your understanding of the argument.

Key Words and Phrases

flow chart   assumed premise   intermediate conclusion

Key Concepts and Skills

- flow charts for arguments with independent premises, additive premises, assumed premises, and intermediate conclusions.
- assessment of soundness or strength of complex arguments.

Unit 1D Solutions

Flow Charts for Arguments.

1. This argument can be analyzed using the four-step method of the book. (Step 1) As shown below, there are two stated premises and one conclusion. (Step 2) The assumed premises are always a matter of judgment (what is obvious to one person, may be a huge assumption to someone else). Three assumed premises are needed to make this argument valid. (Step 3) The layout and flow chart for the argument are as follows:

P1: Asbestos is a known cancer-causing substance.

P2: This school contains asbestos.

A1: The school contains asbestos in amounts that are harmful.

A2: It is important to protect the occupants of the school.

A3: The only way to handle buildings with asbestos is to close them.

C: The school should be closed.

We see that the five premises lead deductively

$$P1 + P2 + A1 + A2 + A3$$
$$\downarrow$$
$$C$$

(additively) to the conclusion. (Step 4). With the assumed premises, the argument is deductively valid. What about its soundness which depends on the truth of the premises? The first premise has been established conclusively. The second premise is a statement of fact. The first assumed premise needs to be verified, and it could be either unambiguous or unverifiable. As with many health issues, there is little agreement on what levels of toxins are safe. The second assumed premise is something most of us could accept. The third assumed premise is also plausible. If the first assumed premise could be verified conclusively, this would be a <u>sound</u> argument.

3. (Step 1) There is one stated premise and one conclusion as shown below. (Step 2) The argument is not complete without an assumed premise to the effect *that if people ride in limousines, then they must be rich*. (Step 3) The argument and its flowchart look like this:

A1: If people ride in limousines, then they must be rich.

<u>P1: I saw Jenny in a limousine.</u>

C: Jenny must be rich.

$$A1 + P1$$
$$\downarrow$$
$$C$$

(Step 4) This deductive argument is <u>valid</u> since it has the form of affirming the antecedent. However, it is <u>not</u> <u>sound</u> since the first premise is false. There are occasions when people who are not rich ride in limousines.

5. (Step 1) There is one stated premise and one conclusion in this argument. (Step 2) An assumed premise is needed to make this argument complete. (Step 3) Here is one way to write the argument and its flowchart.

P1: The U.S. provided weapons to the Afghan rebels.

A1: The Soviet Union lost the war because the Afghan rebels had weapons.

C: The Soviet Union lost the war because the U.S. provide the weapons to the rebels.

$$A1 + P1$$
$$\downarrow$$
$$C$$

(Step 4) This is a deductive argument and it is <u>valid</u>. However, the both premise are arguable. To the extent that one believes the premises, the argument is <u>sound</u>.

7. (Step 1) Each statistic about crime can be considered a stated premise. There is one conclusion. (Step 2) There is an intermediate conclusion to the effect that based on the statistics, there is a crime problem. There must also be an assumed premise stating that increasing the conviction rate and strengthening the police force will decrease crime. (Step 3) Here is one way to present the argument and its flowchart.

P1: A criminal offense occurs every two seconds.

P2: Violent crimes occur every 16 seconds.

P3: Robberies occur every 48 seconds.

I1: There is a crime problem that must be combated.

A1: Increasing the conviction rate and strengthening the <u>police force are the most effective ways to combat crime.</u>

C: We should increase the conviction rate and strengthen the police force

(Step 4) The argument consists of two parts. In the first part, three premises lead independently (inductively) to the intermediate conclusion that there is a crime problem. In the second part of the argument, the intermediate conclusion is combined additively (deductively) with the assumed premise to reach the conclusion. The inductive part of the argument is strong since three facts are given to demonstrate the crime problem. With the assumed premise included, the deductive part of the argument is <u>valid</u>. Politicians and correction officials argue over the best strategy to combat rising crime rates. So the assumed premise is certainly open to debate. To the extent that one believes the assumed premise, the argument is <u>sound</u>.

Quantitative Reasoning.

9. Following the example of the text, let Option A be the $350 option with the $75 cancellation penalty and let Option B be the $600 fully refundable option. If you go on the trip Option A costs $250 less than Option B. If you decide not to go on the trip, Option A costs $75 more than Option B. Since $250 is about three times as much as $75, you might argue that if the likelihood of canceling is more than three times the likelihood of traveling , then Option B is the better choice. Otherwise, take Option A.

11. The Saturday night stay saves $365 on airfare, but the hotel costs an extra $210 for the 2 nights (Saturday and Sunday nights) and meals cost an extra $110 for the 2

days, making the total cost quite similar for the two trip options.

13. First note that a round trip is 3000 miles (the cities are 1500 miles apart). If a free round-trip ticket is awarded with Airline A for every 25,000 miles of travel, then you would have to make the round trip 9 times (because 25,000 mi./3000 mi. = 8.3) before you would get the free round-trip ticket. Therefore, 10 round trips on Airline A costs 9 × $350 = $3150. However, 10 flights on Airline B costs 10 × $325 =$3250. Therefore, if you plan to fly 10 times (or a multiple of 10 times), you should use Airline A. The question becomes more complex if you were to fly more or less than 10 times.

Complete Argument Analysis. It should be said at the outset that there is always more than one way to analyze and interpret complex logical arguments. The following solutions represent just one of many possible approaches. Notice that the four-step method of described in the unit is used, at least implicitly, in each solution.

15. This argument can be written out in the form:

A1: Salaries and funding at a university reflect the institution's priorities.

P1: The athletic department receives more funding than academic programs.

P2: The football coach has a higher salary than people in academic departments.

C: The university values athletics over academics.

In this form the argument consists of one assumed premise, two stated premises and a conclusion. The argument is deductive since the premises must be used additively to reach the conclusion. The argument is valid since the conclusion follows logically from the premises. To the extent that one believes the assumed premise to be true, the argument is sound.

17. This argument can be written out as follows:

P1: In the next decade the number of 5-15 year-old Americans will increase.

P2: In the next decade the number of 25-35 year-old Americans will decrease.

I1: The number of students will increase and the size of the teacher pool will decrease.

C: The demand for teachers will increase in the next decade.

The two stated premises lead to an intermediate conclusion and the conclusion appears to follow directly from the intermediate conclusion (perhaps an assumed premise could be invented to make this step more plausible). This is a valid deductive argument. The two premises are known to be true, so the argument is also sound.

19. There are two stated premises in this argument that lead inductively to an intermediate conclusion (I1) which is then combined with a third stated premise and an assumed premise to give the conclusion. The argument can be written out as follows:

P1: Large numbers of American are uninsured.

P2: Health care costs are skyrocketing.

I1: We have a health care crisis.

P3: Only the government has the power to solve this crisis.

A1: The government can find effective solutions only if people with health care priorities are in office.

C: Support candidates with health care priorities .

The first part of the argument leading to I1 is a fairly strong inductive argument since two persuasive reasons are given to support the claim that there is a health care crisis. The second part of the argument leading to the conclusion is deductive, and it appears to be valid. What about soundness? If we accept I1 as a true statement, then the most questionable premise is P3. There are other solutions to a health care crisis besides government intervention. For example, a private enterprise could be involved in a solution. Although the argument seems solid logically, it fails to be sound because of P3. If the word *only* were removed from this premise, the argument would be more sound.

21. This rather long argument can be distilled into a few lines:

P1: The only way for the many people of China and India to achieve a higher standard of living is to burn their tremendous quantities of domestic coal.

A1: Burning large quantities of coal will lead to an environmental catastrophe.

P2: It is inevitable that China and India will seek a higher standard of living

C: We will have an environmental catastrophe.

This is a deductive argument in which the three premises (one of them assumed) leads to a conclusion. The argument is valid. The argument's soundness depends on whether one believes that the Chinese and Indian people will inevitably seek a higher standard of living. If they do not, then the prediction of the argument is false. One could also question how much coal would need to be burned to create an environmental catastrophe. So the argument is best judged not sound.

23. This argument cites several independent reasons for banning calculators in schools. Thus, it is an inductive argument that can be written as follows:

P1: Pencil and paper methods are best for learning to add fractions.

P2: Pencil and paper methods strengthen mental skills.

P3: Pencil and paper methods (drill) induce good discipline.

P4: The previous generation used only pencil and paper methods

---

C: Fourth graders should not be allowed to use calculators.

Some people would find this inductive argument persuasive. However, it overlooks some of the advantages of calculators. So it might be classified as a strong, but not convincing argument.

25. A Financial Decision. In this problem we must compare the cost of two options for having access to a computer for three months.

a. Expressed as a formal argument there are two stated premises and two intermediate conclusions that lead deductively to a conclusion.

P1: Option A is to lease a computer for 3 months at $350 per month.

P2: Option B is to buy a computer for $2100 plus 5% sales tax and resell it for (an estimated) $1200.

I1: Option A costs $3 \times \$350 = \$1050$.

I2: Option B costs $\$2100 + (0.05 \times \$2100) = \$2205$ to buy the computer; with the $1200 resale, this option costs $\$2205 - \$1200 = \$1005$.

C: Provided all estimates are accurate (including the resale value) the buy/sell option is slightly less expensive.

2. STATISTICAL REASONING

Overview

Much of the quantitative information that flows over us every day is in the form of data that is gathered through surveys and other statistical studies. When a person, business, or organization wants to know something, often the first (and only) idea that comes to mind is "collect data." From news reports to scientific research, from political polls to TV viewer surveys, we are surrounded by statistics. How are of these numbers gathered? Are the conclusions based on those numbers reliable? And most important of all, should you believe the results of a statistical study? Should you change your life based on the results of a statistical study?

These are the questions that we will *begin* to answer in this chapter. The emphasis of this chapter is on rather qualitative issues, and there will not be a lot of computation involved. We will resume out study of statistical problems in Chapter 9 when we look at the quantitative side of statistics. If you master these two chapters, you will have a good foundation in statistical studies — enough to allow you to read the news critically and to take a complete course in statistics.

All in all, this is a fascinating chapter, filled with many practical and real examples. It has an immediate relevance to your studies and your everyday life.

Unit 2A Fundamentals of Statistics

Overview

The goal of this unit is to learn how a statistical study should (and should *not*) be done. Of utmost importance is the distinction between the **population**, the group of people or objects that you would like to learn about, and the **sample**, the group of people and objects that you actually measure or survey. As you will see, there are basically two aspects to any statistical study:

- collecting data from the people or objects in the sample, and
- drawing conclusions about the entire population based on the information gathered from the sample.

As you will see there are many ways that either of these steps can go wrong. Most of this chapter will deal with the first step, which is often called **sampling**. This step is crucial; if a sample is not chosen so that it is representative of the entire population, the conclusions of the study cannot be reliable.

You often hear statements like, "The President's approval rating is 54% with a margin of error of 5%." Results of surveys are often given in this form; for this reason, we will also look briefly at the notions of **confidence interval** and **margin of error**. Although these ideas are rather technical, it is important to understand at least what they mean. In Chapter 9, we will see how confidence intervals and margins or error are actually found.

Finally, statistical studies and surveys take many different forms. We close this unit by pointing out the difference between **observational studies**, in which people or objects are simply queried or measured, and **controlled experiments**, in which researchers actually intervene and create two or more groups of people or objects. Within each of these two categories, there are further important distinctions.

Key Words and Phrases

sample	population	raw data
sample statistics	population parameters	antecedent
random sampling	margin of error	95% confidence interval
observational study	controlled experiment	control group
placebo	single-blind experiment	double-blind experiment
case control study	cases	controls

Key Concepts and Skills

- understand the distinction between a sample and the population.
- know the five basic steps of a statistical study.
- be able to form a sample by simple random sampling.
- understand the difference between an observational study and a case controlled experiment.

Unit 2A Solutions

5.

a. The population is all lawns, or perhaps all lawns in similar climate zones. The sample is the lawns of your 30 friends. The population parameter is the percentage of all lawns that do better with More-Grow.

b. The raw data consist of the 30 responses: 18 for More-Grow and 12 for Go-Green.

c. The sample statistic is the percentage of lawns that do better with More-Grow.

d. Your estimate of the population parameter is 60%. Your friends' lawns are representative of all lawns; furthermore, the sample size is not very large. So the estimate of the population parameter may not be very accurate.

7. Pre-Election Survey. The population is all students who will vote in the election. The sample of students who will vote must be selected randomly, perhaps by a random drawing of student ID numbers, so that there is no bias for age, sex, major, residence (on-campus vs. off-campus). Because most student elections do not register voters, it is difficult to be sure that only students who will vote are sampled. It may be best to ask students if they plan to vote before they are polled, but even this strategy will be subject to some bias.

Samples and Populations.

9. The population is all people in this country. Blood types do not change from birth, so age is not a factor. An unbiased sample must be chosen randomly so it fairly represents the sex and racial background of people in the population.

11. The population is all lung cancer victims. The sample would consist of autopsy records of victims randomly selected to represent the ages, sex, living conditions, and life styles of all victims.

13. The population is presumably all people who are susceptible to colds (isn't everyone?). The random sample should include both people who do drink 3 cups of herbal tea per day and those who don't. The samples must be randomly selected to represent the population in terms of age, sex, living condition, and life style.

Margin of Error in Polls.

15. The 95% confidence interval is 63% − 3.6% to 63% + 3.6% or 59.4% to 66.6%. The confidence interval means that we can be 95% confident that the actual percentage of people who feel that the United States is going through good times is between 59.4% and 66.6%.

percentage of people who agree with the statement is between 68.4% and 71.6%.

17. The 95% confidence interval for the survey is 5.6% − 0.4% to 5.6% + 0.4% or 5.2% to 6.0%. The confidence interval means that we can be 95% confident that the actual percentage of unemployed people is between 5.2% and 6.0%.

Types of Studies.

19. This study is observational. It is a case-control study because the subjects in the sample were chosen to fall naturally into two groups.

21. This study is an observational survey. There are no case-control groups.

23. This study is purely observational. There are no case-control groups.

Types of Statistical Studies.

25. This question would have to be answered by an observational study through the use of a survey. It could not use a placebo, or be single- or double-blind.

Unit 2B Should You Believe a Statistical Study?

Overview

Suppose you read 68% of all TV viewers watched the NCAA basketball final game or that 1 in 5 people in the world do not have access to fresh drinking water, or that a pre-election poll says that the Republican candidate leads by 5 percentage points. In this unit, we address the basic question: how do you know whether to believe the claims of these statistical studies?

The unit takes the form of a list of eight guidelines for evaluating a statistical study. Each guideline is accompanied by examples and analyses. It may not be possible to assess a particular study in light of *all* eight guidelines, but the more of these guidelines that a study satisfies, the more confident you can be about the conclusions of the study. Just for completeness, we'll list the eight guidelines here.

1. Identify the goal, type, and population of the study.

2. Consider the source.

3. Look for bias in the sample.

4. Look for difficulties in *defining* the quantities of interest.

5. Look for difficulties in *measuring* the quantities of interest.

6. If a survey is involved, consider its setting and wording.

7. Check for consistency between results and interpretation.

8. Stand back and consider the conclusions: what does it mean to you?

Key Words and Phrases

bias	peer review	selection bias
participation bias	quantities of interest	availability error

Key Concepts and Skills

- understand the eight guidelines for evaluating a statistical study, and be able to carry them out on a particular study.

Unit 2B Solutions

Bias in Sampling.

12. Unless there were some pattern that mysteriously affected every 100th chip, this process is free of bias.

13. This process is free of bias.

14. While the study could be carried out in an objective fashion, there is opportunity for bias because of the company's financial interest in the outcome.

15. This survey has selection bias because those who go to supermarkets between 10:00 a.m. and noon are not representative of all people or of all beer drinkers. For example, this process might favor unemployed people.

16. Evaluating a Statistical Study.

a. We will agree that "spying on employees" means any of the activities in the table. At most, the sum of all the percentages, or 92%, of all employers surveyed spy on employees (if a different set of employers practice each form of spying).

b. At a minimum, 35% of all employers surveyed spy on employees (if the employers engaged in spying all belong to the 35% category).

c. The conclusion that two-thirds of all employers spy on employees does not follow from the data. Two-thirds or 66.7% is roughly an average of the minimum and maximum figures of parts (a) and (b). But this reasoning is not justified.

d. Guideline 1: Identify the Goal, Type, and Population. None of these points were clarified in the report.

Guideline 2: Consider the Source. While the source (American Management Association) is cited, we don't know if this is an objective group.

Guideline 3: Look for Bias in the Sample. We don't have enough information about the sampling process to evaluate this point.

Guideline 4: Look for Difficulties in *Defining* Quantities of Interest. The quantities of interest (various forms of spying) are well-defined.

Guideline 5: Look for Difficulties in *Measuring* Quantities of Interest. The quantities of interest are certainly difficult to measure. How is evidence of records of employee phone calls gathered: hearsay, hard copies, computer records?

Guideline 6: If a Survey is Involved, Consider its Setting And Wording. We don't have enough information to evaluate this point.

Guideline 7: Check for Consistency Between Results and Interpretation. The study fails badly on this point. How does the two-thirds statement follow from the given data?

Guideline 8: Stand Back and Consider the Conclusions: The study has an alarmist tone. Nevertheless, it suggests that some concern about employer surveillance may be warranted.

Unit 2C Basic Statistical Graphs

Overview

We often think of data and statistics as long lists of numbers, and indeed they start in that form. But these numbers don't become meaningful until they are summarized in some digestible form; and one of the best ways to represent long lists of numbers is with a picture. In this and the next unit, we will explore the many ways in which quantitative information can be displayed. This unit focuses on the basic types of graphs, ones that can often be drawn with a pencil and paper. The next unit highlights the more exotic types of graphs which are more difficult to produce, but equally common in news and research reports.

Here is a summary of the types of graphs that we will consider in this unit.

- **Bar graphs** are used to show how some numerical quantity (for example, population) varies from one category to another (for example, for several different countries). The categories are usually non-numerical and can be shown in any order.

- **Pie charts** are used to show the fraction (or percentage) of a population that falls into various categories. The categories are usually non-numerical and can be shown in any order. For example, a pie chart could be used to show the fraction of a class that has brown, black, blond, and red hair.

- **Histograms** are bar graphs in which the categories are numerical, and thus have a natural order. Histograms often show how many objects or people are in various categories. For example, a histogram would be used to show the number of people in a town that fall into age categories 0–9, 10–19, 20–29, and so on.

- **Line charts** serve the same purpose as histograms, but instead of using bars to indicate the number of objects or people in each category, they use dots connected by lines.

- **Time-series diagrams** are usually histograms of line charts that show how a quantity changes in time. For example, the day-to-day changes in the stock market would be displayed as a time-series diagram.

- **Scatter plots** are used to show how *two* quantities for each of many objects or people are related. For example, if you measured the height and weight of each student in your class, you could display the data with a scatter plot.

We might point out one possible source of confusion. There seems to be no standard definition of histograms and bar graphs. We have tried to choose a distinction that is common in many books: a histogram is any graphs that uses bars, while a histogram is a particular kind of bar graph that is used for ordered quantitative data.

Key Words and Phrases

bar graph	pie chart	histogram
line chart	time-series diagram	scatter plot
correlation		

Key Concepts and Skills

- determine an appropriate kind of display for a given set of data.
- display an appropriate set of data with a bar graph by finding the correct height of the bars.
- display an appropriate set of data with a pie chart by finding the correct angles for the sectors of the pie.
- display an appropriate set of data with a histogram by finding the correct height of the bars.
- display an appropriate set of data with a line chart by finding the correct location of points on the line chart.
- display an appropriate set of data with a time-series diagram by finding the correct location of points on the diagram.
- display an appropriate set of data with a scatter plot by finding the correct location of points on the plot.

Unit 2C Solutions

1. Bar Graph Analysis.

a. The ratio of Asians to African is approximately 270,000/45,000 = 6; that is 6 times as many Asians immigrated than Africans.

b. The height of the bar graph corresponds to 300,000. Therefore the height of the bar for Russia would be 54,494/300,000 = 18% of the height of the bar graph, which is 18% × 6 cm = 1.1 cm. The bar for South America would be 45,666/300,000 = 15% of the height of the bar graph, which is 15% × 6 cm = 0.9 cm.

3. Income by Education.

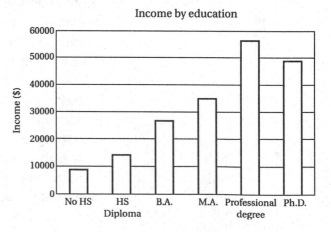

5. Pie Chart of Language.

a. The number of people in the "All Other" category is
$31.2 - (17.3 + 1.7 + 1.5 + 1.3 + 1.2) = 8.2$ (million)
people.

b. The percentages for each language are as follows.
Spanish: $17.3/31.2 = 55.4\%$, French: $1.7/31.2 = 5.5\%$,
German: $1.5/31.2 = 4.8\%$, Italian: $1.3/31.2 = 4.2\%$,
Chinese: $1.2/31.2 = 3.8\%$, All Others: $8.2/31.2 =$
26.3%. The angle for a particular category is that
category's percentage of $360°$: Spanish: $360° \times 55.4\%$
$= 199°$, French: $360° \times 5.5\% = 20°$, German: $360° \times$
$4.8\% = 17°$, Italian: $360° \times 4.2\% = 15°$, Chinese: $360°$
$\times 3.8\% = 14°$, All Others: $360° \times 26.3\% = 95°$.

c.

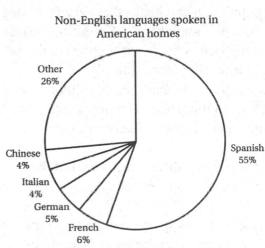

Non-English languages spoken in
American homes

7. Women Workers.

a, b.

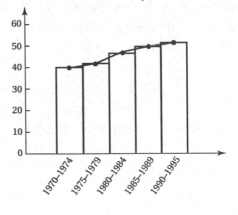

Percent of women workers
with full-time jobs

c. Clearly there is an upward trend in the data, as the
percentage of percentage of full-time women workers
increases with time.

9. Voter Turnout.

a. Voter turnout was lowest in 1996 (54.2% of eligible
voters actually voted).

b. Unemployment was highest in 1976 (7.7%).

c. Visually there does not seem to be a correlation
between voter turnout and unemployment.

d.

Correlations.

13. Age and height of adults are uncorrelated.

15. We might discover a mild (negative) correlation in
which countries with larger family sizes also have smaller
life expectancies.

Unit 2D Graphics in the Media

Overview

Whereas the previous unit dealt with graphs that are relatively easy to produce yourself, this unit will survey more complicated displays that appear frequently in the media. The goal in this unit is not to actually produce these more sophisticated graphs; often powerful software packages are needed to create them. Rather the emphasis will be on interpreting these displays of quantitative information.

Here is a summary of the types of graphs that we will consider in this unit.

- **Multiple bar graphs** are used to show how *two or more* numerical quantities (for example, population and birth rate) vary from one category to another (for example, for several different countries). Multiple bar graphs are really two or more bar graphs combined with each other; they include multiple histograms.
- **Stack plots** are used to display several quantities simultaneously often showing how they all change in time. For example, a stack plot could be used to show how the incidence of several diseases have changed in time.
- **Three-dimensional** graphs take many different forms, but they are all used to show how three different quantities (or variables) are related to each other.
- **Contour plots**, such topographic maps, give a two-dimensional picture of how one quantity varies over a geographical region. Weather maps showing temperature or pressure often take the form of contour plots.

As graphical displays become more complex, there is more opportunity for confusion and deception. The unit closes by discussing several ways in which you can be misled by such displays. Although graphs can be beautiful and effective, they must be interpreted with caution!

Key Words and Phrases

multiple bar graph	stack plot	three-dimensional graphics
contour plots	inflation adjusted data	exponential scale
pictograph		

Key Concepts and Skills

- interpret multiple bar graphs and histograms
- interpret stack plots

- interpret three-dimensional graphics
- interpret contour plots
- detect deceptive pie charts
- understand inflation adjustments for data
- interpret percent change graphs
- interpret exponential scales
- detect deceptive pictographs.

Solutions Unit 2D

24. Net Grain Production.

a. Only the difference between grain production and grain consumption is shown on the graph.

b. Brazil's net grain production for 1990 was 37 million tons – 43 million tons = – 6 million tons. The bar for this data point would extend downward about one-eighth of the way to the –50 mark.

25. Government Outlays Stack Plot.

a., b. We must measure the width of the band labeled *net interest* and *national defense* at each year and then convert that width to percent of total outlays. The figures will be approximations.

Year	Net Interest	Defense
1960	8%	52%
1970	7%	42%
1980	8%	22%
1990	15%	24%
1999	14%	15%

c. For 1988, we can construct the following table from the stack plot. The amount spent on each category is the corresponding percentage of $1 trillion.

	Percent of Budget	Amount Spent ($ billions)
Individuals	50%	500
Defense	27%	270
Interest	14%	140
Other	9%	90

d. For 1995, we can construct the following table from the stack plot. The amount spent on each category is the corresponding percentage of $1.5 trillion.

	Percent of Budget	Amount Spent ($ billions)
Individuals	60%	900
Defense	20%	300
Interest	12%	180
Other	8%	120

26. Three-Dimensional Bird Graph.

a. The labels 1, 2, ..., 8 give hours after 8:30 P.M. So they correspond to 9:30 P.M., 10:30 P.M., ..., 4:30 A.M.

c. Looking at the curve for Oneonta, at one hour after 8:30 PM, it appears that approximately 10 birds passed over Oneonta. At 11:30 P.M. (3 hours after 8:30 P.M.), perhaps 35 birds passed over Oneonta. And at 1:30 A.M. (5 hours after 8:30 P.M.), perhaps 60 birds passed over Oneonta.

e. To estimate the total number of birds over a city during the entire evening, we must add the bird counts for each hour (another way to look at the procedure is that we must estimate the area under each curve). It appears that Oneonta and Jefferson had the most sightings.

27. Tuition Increases.

a. Tuition did not decrease in any of the categories because all categories showed (positive) increases in tuition.

c. In 1988 private college tuition rose 9% while the consumer price index rose 4%. This year showed the greatest difference.

e. In 1990 private college tuition rose about 7.5%. Therefore tuition rose about 7.5% × $10,000 =$750. This means the tuition in 1991 would have been about $10,750.

28. U.S. Suicide Rates.

a.

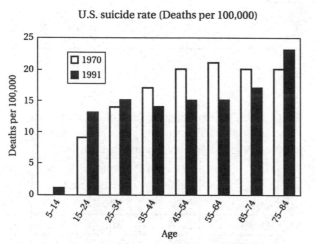

U.S. suicide rate (Deaths per 100,000)

b. Overall, suicide rates tend to increase with age. Among the younger (less than 35) and oldest (over 75) age groups, the suicide rate increased between 1970 and 1991. Among the middle age groups, the suicide rate decreased between 1970 and 1991.

29. Distorted Pies. The proportions appear to be about Army 65%, Navy 30%, and Marines 5%. The actual percentages are 70%, 26%, and 4%. This means that Army casualties were 70% × 16,112,566 = 11,278,796; Navy casualties were 26% × 16,112,566 = 4,189,267; and Marine casualties were 4% × 16,112,566 = 644,503. A flat pie chart would be easier to read accurately. It would help if the percentages appeared on the pie chart.

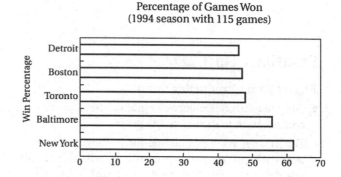

Percentage of Games Won
(1994 season with 115 games)

Unit 2E Causal Connections

Overview

In Unit 2C, we explored scatter plots and discussed how they can be used to identify **correlations** between two quantities. This unit returns to the idea of correlations and opens with the claim that when two quantities appear to correlated, it may be due to

- a coincidence,
- a common underlying cause, or
- a genuine cause and effect relation.

Of the three possibilities, the most interesting and the most difficult to establish is the last one: when can we be sure that one event causes another?

The first step in discussing cause and effect relations is to understand three different types of cause:

- A cause is **necessary** if the effect cannot happen in its absence.
- A cause is **sufficient** if the effect always occurs in its presence.
- A cause is **probabilistic** if its presence increases the likelihood of the effect.

The unit becomes a bit philosophical at this point by presenting four methods attributed to John Stuart Mill for establishing cause and effect relations. These methods are

- the method of agreement,
- the method of difference,
- the method of concomitant variation, and
- the method of residues.

Because causality is important ways in legal cases, we also look at the legal standards for establishing causality. They are crucial in determining the guilt or innocence of defendants. They are

- possible cause,
- probable cause, and
- cause beyond reasonable doubt.

The unit closes by presenting four different (real) case studies in which identifying a cause and effect relation led to a discovery or the solving of a mystery.

Key Words and Phrases

necessary cause	sufficient cause	probabilistic cause
method of agreement	method of difference	method of residues
method of concomitant variation	cause beyond reasonable doubt	
probable cause	possible cause	

Key Concepts and Skills

- identify each of the three explanations for a correlation.
- identify each of the three types of cause.
- apply Mill's methods.
- understand the three legal levels of confidence in causality.

Unit 2E Solutions

Interpreting Correlations.

30. One could argue that there is a <u>direct causal link</u> in this correlation: the increased number of crimes committed per day causes an increase in the number of prisoners. However, this assumes that the law enforcement and court systems are capable of increasing the number of convictions. One could also argue that there is an <u>underlying cause</u> in this correlation that is the common cause of both effects. The most likely factor is the increasing population: as the population increases, one would expect both the crime rate and the number of prisoners to increase.

31. This is a case of neither coincidence nor direct causal link, but rather an <u>underlying cause</u> for both phenomena. The underlying cause is the increase in the population of Los Angeles. More freeways are built to serve an increasing population and an increasing population means more cars, buses, and trucks (to service the increasing population) on the roadways. A secondary effect is that with more and larger freeways people change their driving habits and tend to drive more.

32. A correlation between a student's grade and the age of her mother would have to be considered a <u>coincidence</u>.

33. There is a plausible <u>causal link</u> between gasoline prices and attendance in National Parks. As gasoline prices rise, fewer people would choose to take driving vacations.

34. A ban on deer hunting would tend to increase the deer population which would attract lions that prey on deer. There is a <u>direct causal link</u>.

35. As merchants and customers know, as the price of an item increases, the demand (or number of items sold) decreases. There is a direct <u>causal link</u>.

Causal Conditions.

36. Getting caught for speeding is a <u>sufficient cause</u> for paying a large fine (but not necessary since one can pay a large fine for other reasons).

37. Being a famous movie star does not guarantee that you will get a certain role (although it improves your chances). And if you get a certain role, it does not always mean you are a famous star (although it is often the case). So the connection here is a <u>probabilistic</u> cause. **38.** Winter is a <u>probabilistic cause</u> of cold weather. It is possible to have cold weather in seasons other than winter and warm weather can occur in the winter.

39. It is now believed that ozone depletion in the atmosphere causes an increase in skin cancer. Therefore, when considering many people, ozone depletion is a <u>sufficient cause</u> of skin cancer.

40. The connection here is a <u>probabilistic cause</u>. Exposure to a flu-infected person does not guarantee contraction of flu (although it may be likely) and there are many other ways to contract flu.

41. Probabilistic causes are the rule in health matters and there is as much uncertainty in economics as in health! While rising interest rates are tied closely to declines in the stock market, the cause is at best <u>probabilistic</u>.

42. Poor mathematical training is a <u>probabilistic cause</u> for failing to land managerial positions! It is not a necessary condition because there are other ways to be rejected from managerial positions.

3. PROBLEM SOLVING TOOLS

Overview

The first two chapters of the book were devoted to *qualitative* issues — topics that don't require extensive use of numbers and computation. In this chapter (and the remainder of the book) we turn to *quantitative* matters. Perhaps it's not surprising that we begin our study of quantitative topics with problem solving. The first two units of the chapter deal with a very basic and important problem solving technique, the use of units. The last unit of the chapter is an overview of various problem solving strategies. It is a valuable chapter whose lessons run through the rest of the book.

Unit 3A Using Unit Analysis

Overview

Nearly every number that you encounter in the real world is a measure of *something*: 6 billion *people*, 5280 *feet*, 5 trillion *dollars*, 26 *cubic feet*. The quantities that go with numbers are called **units**. And the message of this unit of the book is that using units can simplify problem solving immensely. Indeed the use of units is one of the most basic problem solving tools.

We begin by considering the most basic **simple units** for the fundamental types of measurement. Here are a few examples of simple units.

- length — inches, feet, meters
- weight — pounds, grams
- capacity — quarts, gallons, liters
- time — seconds, hours.

From these simple units, we can build endless **compound units**. The rest of the unit is spent illustrating a wide variety of such compound units. Among the many you will meet and work with are units of

- units of area such as square feet and square yards
- units of volume such as cubic inches and cubic feet
- units of speed such as miles per hour
- units of price such as dollars per pound
- units of gas mileage such as miles per gallon.

One of the realities of life is that there are many units for the same quantity. For example, we can measure lengths in inches, centimeters, feet, meters, miles, or kilometers. For example, we measure rooms in square feet, but have to buy carpet in square yards. Thus one of the necessities of problem solving is being able to convert from one unit to another consistent unit. The key to doing conversions between units is to realize that there are three equivalent ways to express a conversion factor. For example, we can say 1 foot = 12 inches, or we can say 1 foot *per* 12 inches, or we can say 12 inches *per* foot. Mathematically we can write

$$1 \text{ foot } = 12 \text{ inches } \quad \text{or} \quad \frac{1 \text{ foot}}{12 \text{ inches}} \quad \text{or} \quad \frac{12 \text{ inches}}{1 \text{ foot}}.$$

These three forms of the same conversion factor are absolutely equivalent. The key to a happy life with units is choosing the appropriate form of the conversion factor for a given situation.

Additional problem solving skills can then be built on these fundamental skills. You will see and learn how to make a chain of conversions factors to solve more complex problems. For example, the time required to count a billion dollars at the rate of one dollar per second is

$$\$1,000,000,000 \times \frac{1 \text{ sec}}{\$1} \times \frac{1 \text{ min}}{60 \text{ sec}} \times \frac{1 \text{ hr}}{60 \text{ min}} \times \frac{1 \text{ day}}{24 \text{ hr}} \times \frac{1 \text{ yr}}{365 \text{ days}} = 31.7 \text{ years.}$$

The fact that the units cancel and give an answer in *years* tells you that the problem has been set up correctly.

Finally we look at a very practical type of unit conversion problem, those associated with currency. If you travel to France you will quickly learn that 1 franc is equal to about 18 cents. From this fact, you will also want to answer questions such as

- which is larger, 1 franc or 1 dollar?
- how many francs in a dollar?
- how many dollars in a franc?
- how many dollars in a 23.45 francs?
- if apples cost 23 francs per kilogram, what is the price in dollars per pound?

Here are a few final words of advice: Never was the motto *practice makes perfect* more true than with problem solving and units. You should work all assigned problems, and then some, in order to master these techniques. And unless your instructor tells you otherwise, there is no need to memorize hundreds of conversion factors. It's helpful to know a few essential conversion factors off the top of your head. As for the rest, it's easiest just to know how to find them quickly in the book.

Key Words and Phrases

simple units	compound units	area
Volume	conversion factor	

Key Concepts and Skills

- convert from one simple unit to another; for example, from inches to yards.
- convert from one unit of area to another unit of area; for example, from square inches to square yards.
- convert from one unit of volume to another unit of area; for example, from cubic inches to cubic yards.
- solve problems involving chains of conversion factors; for example, finding the number of seconds in a year.
- convert from one unit of currency to another; for example, from dollars to pesos.

Unit 3A Solutions

1. Identifying Compound Units.

a. A liter holds 1000 cubic centimeters means there are 1000 cm³/liter.

b. One atmosphere is 14.7 pounds per square inch means one atmosphere is 14.7 pounds/in².

c. The Earth's average density is 5.5 grams per cubic centimeter 5.5 grams/cm³.

d. The rocket accelerated at 7 meters per second per second or 7 meters/ sec².

e. The ore contains 5 ounces of gold in every cubic foot or 5 ounces/ft³ of gold.

2. U.S. Areas – Colorado.

a. Recall that the area of a rectangle is the product of its length and its height (length × height). If the United States is viewed as a rectangle with length and height 3500 miles and 1000 miles, respectively, its area is (3500 miles) × (1000 miles) = 3,500,000 square miles. Recall that to find the conversion factor from square miles to square feet we can square (raise to the second power) the conversion factor 1 mile = 5280 ft. This results in

$$(1 \text{ mile })^2 = (5280 \text{ ft})^2 \text{ or } 1 \text{ mi}^2 = 27,878,400 \text{ ft}^2.$$

Using this conversion factor, we find that 3,500,000 square miles is equal to

$$3,500,000 \text{ mi}^2 \times 27,878,400 \frac{\text{ft}^2}{\text{mi}^2} = 97,574,400,000,000 \text{ ft}^2.$$

(We will soon develop scientific notation which avoids writing out so many zeros!) The conversion factor for square mile to acres is 1 mi² = 640 acres. Thus, the area of the United states is equal to

$$3,500,000 \text{ mi}^2 \times 640 \frac{\text{acre}}{\text{mi}^2} = 2,240,000,000 \text{ acres}.$$

b. Calculations similar to those in part (a) give the area of Colorado in various units. Assuming that Colorado is a rectangle with length and height of 400 miles and 300 miles, respectively, the area of the state is (400 miles) × (300 miles) = 120,000 mi². The area in square feet is

$$120,000 \text{ mi}^2 \times 27,878,400 \frac{\text{ft}^2}{\text{mi}^2} = 3,345,408,000,000 \text{ ft}^2.$$

The area of Colorado in acres is

$$120,000 \text{ mi}^2 \times 640 \frac{\text{acre}}{\text{mi}^2} = 76,800,000 \text{ acres}.$$

c. The fraction of the total area of the United States comprised by Colorado is

$$\frac{120,000 \text{ mi}^2}{3,500,000 \text{ mi}^2} = 0.034 = 3.4\%.$$

Note that we could have taken the ratio of areas in *any* units (for example, acres or square feet).

3. Units of Volume. Recall that the volume of a box is the product of its length, width, and height (length × width × height).

a. The volume of the swimming pool is

(3 meters) × (10 meters) × (5 meters) = 150 meter^3.

b. The volume of the package is (22 in) × (15 in) × (12 in) = 3960 in^3.

c. Notice that the area of the base of the box is already given. To obtain the volume the area of the base must be multiplied by the height of the box. The volume of the skyscraper is (1000 ft) × (25,000 ft^2) = 25,000,000 ft^3.

4. Working with Units. In all of these problems, notice how the units guide the calculation and provide a check on the final answer. All numbers should have units attached to them!

a. The total amount paid is

$$3 \text{ pounds} \times \frac{\$0.50}{\text{pound}} = \$1.50.$$

You will pay $1.50 for the apples.

b. The number of centimeters in a kilometer is

$$\frac{100 \text{ cm}}{\text{meter}} \times \frac{1000 \text{ meter}}{\text{km}} = 100,000 \frac{\text{cm}}{\text{km}}.$$

There are 100,000 centimeters in a kilometer.

c. The number of yards in a mile is

$$\frac{5280 \text{ ft}}{\text{mile}} \times \frac{1 \text{ yd}}{3 \text{ ft}} = 1760 \frac{\text{yd}}{\text{mile}}.$$

There are 1760 yards in a mile.

d. The weight of the ruby in carats is

$$252 \text{ mg} \times \frac{1 \text{ carat}}{200 \text{ mg}} = 1.26 \text{ carats}.$$

The ruby weighs 1.26 carats.

Unit Analysis: What Went Wrong?

5. Sorry! You are wrong because your answer has units of lb^2/dollar. The correct answer is $7.70/lb × 0.11 pound = $0.85.

6. Sorry! The units of your answer are lb/dollar not dollars/lb. The correct answer is $11/50 lb = $0.22/lb. The price per pound of the 50-pound bag is less than the price per pound of the 1-pound bag.

7. Unit Conversions.

a. The number of ounces in a gallon is

$$\frac{8 \text{ oz}}{\text{cup}} \times \frac{4 \text{ cups}}{\text{qt}} \times \frac{4 \text{ qt}}{\text{gal}} = 128 \frac{\text{oz}}{\text{gal}}.$$

There are 128 ounces per gallon.

b. The number of minutes in a year is

$$\frac{60 \text{ min}}{\text{hr}} \times \frac{24 \text{ hr}}{\text{day}} \times \frac{365 \text{ day}}{\text{yr}} = 525,600 \frac{\text{min}}{\text{yr}}.$$

There are 525,600 minutes in a year.

c. Using the fact that 1 hour = 3600 seconds, a speed of 55 miles per hour is the same as

$$\frac{55 \text{ mile}}{\text{hr}} \times \frac{5280 \text{ ft}}{\text{mile}} = 290,400 \frac{\text{ft}}{\text{hr}} \quad \text{and}$$

$$\frac{55 \text{ mile}}{\text{hr}} \times \frac{5280 \text{ ft}}{\text{mile}} \times \frac{1 \text{ hr}}{3600 \text{ sec}} = 80.7 \frac{\text{ft}}{\text{sec}}.$$

We see that 55 miles per hour is 290,400 feet per hour or 80.7 feet per second.

8. Conversions with Units Raised to Powers.

a. Since 1 ft = 12 in, we have that (1 ft)^2 = (12 in)^2 which means that 1 ft^2 = 144 in^2.

b. The area of a football field is (100 yd) × (55 yd) = 5500 yd^2. To express this area in square feet, we must use the conversion factor 1 yd^2 = 9 ft^2. The area of the football field is also equal to

$$5500 \text{ yd}^2 \times \frac{9 \text{ ft}^2}{\text{yd}^2} = 49,500 \text{ ft}^2 \quad \text{and}$$

$$5500 \text{ yd}^2 \times \frac{9 \text{ ft}^2}{\text{yd}^2} \times \frac{1 \text{ acre}}{43,560 \text{ ft}^2} = 1.14 \text{ acres}.$$

A football field has an area of 5500 yd^2 or 49,500 ft^2 or 1.14 acres.

c. Starting with 1 yd = 3 ft, we can cube both sides and find that 1 yd^3 = (3 ft)^3 = 27 ft^3. Therefore 5 yd^3 is the same as

$$5 \text{ yd}^3 \times \frac{27 \text{ ft}^3}{1 \text{ yd}^3} = 135 \text{ ft}^3.$$

9. New Flooring. A problem like this is not unusual! It is not uncommon to find floor tile sold by the square foot and carpet sold by the square yard. For each room, we must multiply the area of the room by the price of the floor covering using compatible units. The cost of the kitchen floor is

$$12 \text{ ft} \times 10 \text{ ft} \times \frac{\$4}{\text{ft}^2} = \$480.$$

The cost of the bedroom floor, after converting to square yards, is

$$14 \text{ ft} \times 15 \text{ ft} \times \frac{1 \text{ yd}^2}{9 \text{ ft}^2} \times \frac{\$14.25}{\text{yd}^2} = \$332.50.$$

The cost of the living room, after converting to square yards, is

$$16 \text{ ft} \times 12 \text{ ft} \times \frac{1 \text{ yd}^2}{9 \text{ ft}^2} \times \frac{\$18}{\text{yd}^2} = \$384.$$

The cost of the bathroom is

$$3 \text{ ft} \times 5 \text{ ft} \times \frac{\$7}{\text{ft}^2} = \$105.$$

And the cost of the hallway is

$$3 \text{ ft} \times 8 \text{ ft} \times \frac{1 \text{ yd}^2}{9 \text{ ft}^2} \times \frac{\$18}{\text{yd}^2} = \$48.$$

Adding up the cost of the five rooms, we find a total of $1349.50 for all of the floor covering.

10. Monetery Conversions II.

a. We are given that $1= 1580 lire. Because it takes more than 1 lire to make $1, it follows that 1 lire is less that $1.

b. Use the conversion factor 1 mark = $0.72:

$$17 \text{ marks} \times \frac{\$0.72}{1 \text{ mark}} = \$12.24.$$

We see that 17 marks is worth $12.24.

c. Use the conversion factor 1 pound = $1.60:

$$5.25 \text{ pounds} \times \frac{\$1.60}{1 \text{ pound}} = \$8.40.$$

We see that 5.25 pounds is worth $8.40.

d. We are given that $1= 1580 lire. Therefore

$$\$19.95 \times \frac{1580 \text{ lire}}{\$1} = 31{,}521 \text{ lire}.$$

We see that $19.95 is worth 31,521 lire.

e. It is easiest to convert marks to francs by way of dollars:

$$200 \text{ marks} \times \frac{\$0.72}{1 \text{ mark}} \times \frac{1 \text{ franc}}{\$0.21} = 686 \text{ francs}.$$

We see that 200 marks is worth 686 francs.

11. Dog Years. We are really using a conversion factor in this problem: 1 real year = 7 dog years.

a. The conversion factor can be used to find that 15 real years corresponds to

$$15 \text{ real years} \times \frac{7 \text{ dog years}}{1 \text{ real year}} = 105 \text{ dog years}.$$

We discover that 15 years in our lifetime is like 105 years in a dog's life.

b. Since 1 dog year = 1/7 real years, the third year of a dog's life begins at 2/7 real years and lasts to 3/7 real years which is from about 15 weeks to 22 weeks.

c. The idea of a dog year comes from scaling the average life time of a human being (say 70 years) down to the average lifetime of a dog (say, 10 years). As a vet can affirm, the relationship is not exact because dogs have different growth patterns and stages than humans.

12. Computer Stored Books. The goal is to estimate the number of pages of text that can be stored on a 500-megabyte hard drive. We need to make several assumptions or estimates. As stated in the problem, one byte codes for one character. Let's also assume that the average word is five characters long and that the average page has 400 words. Then noting that a 500-megabyte hard drive can hold 500 million = 500,000,000 bytes, we can use the units to find that the number of pages on the hard drive is

$$500{,}000{,}000 \text{ bytes} \times \frac{1 \text{ char}}{1 \text{ byte}} \times \frac{1 \text{ word}}{5 \text{ char}} \times \frac{1 \text{ page}}{400 \text{ words}} = 250{,}000 \text{ pages}.$$

We see that a 500-megabyte hard drive can hold about 250,000 pages of text. If this estimate seems high, we might want to revise (upwards) our estimates of characters per word and words per page.

13. The Glen Canyon Dam. Scientific notation is useful in this problem. But since it has not been covered in the text, we will write out all numbers in full.

a. The total amount of water released in a week is

$$25{,}800 \frac{\text{ft}^3}{\text{sec}} \times 60 \frac{\text{sec}}{\text{min}} \times 60 \frac{\text{min}}{\text{hr}} \times 24 \frac{\text{hr}}{\text{day}} \times$$
$$7 \frac{\text{day}}{\text{wk}} = 15{,}603{,}840{,}000 \text{ ft}^3.$$

b. The fraction represented by the water released is again

$$\frac{15{,}600{,}000{,}000 \text{ ft}^3}{1{,}200{,}000{,}000{,}000 \text{ ft}^3} = 0.013 = 1.3\%.$$

The amount of water relaesed during the spike flood was 1.3% of the total water in the reservoir. It seems like a relatively small amount.

Unit 3B Systems of Standardized Units

Overview

In this unit we stay with the themes of units and conversion factors, but now look more deeply into the two standard systems of units: The U.S. Customary System of Measurement (or USCS system, which is used primarily in the Unites States) and the metric system (which is used everywhere else in the world).

We first proceed systematically and survey the USCS units for length, weight, and capacity. Tables 3.2, 3.3, and 3.4 contain many conversion factors, but you should focus on *using* these conversion factors, not memorizing them! Having seen the complications of the USCS system, the metric system should come as a welcome relief. Next we present the metric units for length, weight and capacity. As you will see, the system is based on powers of ten and standard prefixes, which makes conversions between units relatively simple.

Unfortunately, for people living in the United States, conversions between the metric system and the USCS system are necessary. (If you would like to avoid doing such problems, then you should work on getting the United States to go metric!) Table 3.7 has a few of the essential conversion factors between the metric and USCS systems. You may want to write other useful conversion factors in the margin near this table.

Armed with all of these units and conversion factors (and don't forget currency conversion factors as well), we can do even more elaborate problem solving; this is the goal of the remainder of the unit and the problems at the end of the unit.

Key Words and Phrases

USCS system	metric system	meter
gram	liter	second
metric prefixes	Celsius	Fahrenheit
Kelvin		

Key Concepts and Skills

- given the required conversion factor, convert between two consistent USCS units; for example, rods to miles.
- multiply and divide powers of ten.
- know basic metric units and commonly used prefixes.
- given the required conversion factor, convert between two consistent metric units; for example, millimeters to kilometers.

- given the required conversion factor, convert between a USCS unit and a consistent metric unit; for example, ounces to liters.
- convert between temperatures in the three standard systems.
- solve problems using chains of conversion factors involving USCS units, metric units, and currency units.

Unit 3B Solutions

3. USCS Units.

a. Using the conversion factor 1 ft = 12 in, a height of 5'10'' is (5 ft × 12 in/ft) + 10 in = 60 in + 10 in = 70 in. Therefore, 5'10'' is the same as 70 inches.

b. Using the conversion factors 1 ft = 12 in and 1 yd = 3 ft, we see that 6.75 ft = 6.75 ft × 12 in/ft = 81 in. Similarly, 6.75 ft = 6.75 ft × (1 yd/3 ft) = 2.25 yd.

c. The race is 10.2 furlongs in length. The required conversion factor is 1 mile = 8 furlongs. Therefore, the length is

$$10.2 \text{ furlongs} \times \frac{1 \text{ miles}}{8 \text{ furlongs}} = 1.275 \text{ miles.}$$

We find that 10.2 furlongs equals 1.275 miles which is 4.9 miles shorter than a 6.2-mile road race and 20.4 miles shorter than a 26-mile marathon.

d. The relevant conversion factor is 1 pound = 16 oz av. Therefore, a gallon of water weighs about

$$128 \text{ oz av} \times \frac{1 \text{ pound}}{16 \text{ oz av}} = 8 \text{ pounds.}$$

Using a similar calculation, the *exact* weight of a gallon of water is 8.31 pounds.

e. The conversion factor 1 oz av = 437.5 grains (Table 3.3) is needed. Then we find that a letter weighing 154 grains also weighs

$$154 \text{ grains} \times \frac{1 \text{ oz av}}{437.5 \text{ grains}} = 0.352 \text{ oz av.}$$

A letter weighing up to about a third of an ounce can be sent at the lowest rate.

f. The conversion to liquid pints is straightforward. Because 1 qt = 2 pints (liquid) and 1 gallon = 4 quarts, it follows that 0.5 gallons = 4 liquid pints. From Table 3.4, we see that 1 quart has a volume of 57.75 in³.

Therefore 0.5 gallons = 2 quarts = 115.5 in³. The same table tells us that 1 dry pint = 33.6 in³. Therefore

$$0.5 \text{ gal} = 115.5 \text{ in}^3 \times \frac{1 \text{ dry pint}}{33.6 \text{ in}^3} = 3.44 \text{ dry pints.}$$

A half-gallon container holds 4 liquid pints and 3.44 dry pints.

g. From Table 3.4, 1 bushel = 32 dry quarts. Because 1 dry quart has a volume of 67.2 in³, it follows that 1 bushel = 2150.4 in³. With this fact we can convert 5000 and 150 million bushels to cubic inches:

$$5000 \text{ bushels} \times \frac{2150.4 \text{ cu in}}{\text{bushel}} = 10,752,000 \text{ cu in.}$$

$$150 \text{ million bushels} \times \frac{2150.4 \text{ cu in}}{\text{bushel}} = 322,560 \text{ million cu in.}$$

Recall from Unit 3A that 1 ft³ = 12³ in³ = 1728 in³. We can now continue the conversion above from bushels to cubic inches and convert 150 million bushels to cubic feet:

$$322,560 \text{ mill cu in} \times \frac{1 \text{ cu ft}}{1728 \text{ cu in}} = 187 \text{ million cu ft.}$$

A building with a volume of 5000 bushels or almost 11 million cubic inches has a volume of about 6350 cubic feet. If this building has a ceiling height of 8 feet, it would have a floor area of almost 800 square feet.

5. Metric Prefixes. Table 3.6 is very helpful for this problem.

a. We know that 1000 mm = 1 m. So the ratio of 1 mm to 1 m is 1 mm/1000m = 0.001. A meter is 1000 times larger than a millimeter or a millimeter is 1/1000 of a meter.

b. We know that 1000 gm = 1 kg. So the ratio of 1 gm to 1 kg is 1 gm/1000 gm = 0.001. A kilogram is 1000 times larger than a gram or a gram is 1/1000 of a kilogram.

c. We need the facts that 1 milliliter = 0.001 liters and 1 deciliter = 0.1 liters. Therefore, the ratio of 1 milliliter to 1 deciliter is 0.001 liter/0.1 liter = 0.01 which means a milliliter is one one-hundredth of a deciliter.

d. Note that 1 km = 1000 m and 1 micrometer = 0.000001 m. Therefore, the ratio of 1 km to 1 micrometer is 1000 m/0.000001 m = 1,000,000,000 which means a kilometer is one billion times larger than a micrometer.

7. USCS – SI Conversions.

a. The required conversion factors are 1 foot = 12 inches and 1 inch = 2.54 cm = 0.0254 meters. This means   10 m is

$$10 \text{ m} \times \frac{1 \text{ in}}{0.0254 \text{ m}} \times \frac{1 \text{ ft}}{12 \text{ in}} = 32.808 \text{ ft.}$$

We see that 10 meters is the same as 32.808 feet.

b. The relevant conversion factor is 1 m = 1.094 yd. Therefore, 880 yards is equal to

$$880 \text{ yd} \times \frac{1 \text{ m}}{1.094 \text{ yd}} \times \frac{1 \text{ km}}{1000 \text{ m}} = 0.80 \text{ km.}$$

We see that 880 yards is 0.8 kilometers or 800 meters.

c. The relevant conversion factor is 1 gallon = 3.785 liters. Therefore, 20 gallons is equal to

$$20 \text{ gal} \times \frac{3.785 \text{ l}}{\text{gal}} = 75.7 \text{ l.}$$

We see that 20 gallons is about 75 liters.

d. We must note that 1 ml = 1 cm³ and 1 in = 2.54 cm. It then follows that $(1 \text{ in})^3 = (2.54 \text{ cm})^3 = 16.4 \text{ cm}^3$; or 1 in³ = 16.4 cm³. Therefore, 5 milliliters is equal to

$$5 \text{ ml} \times \frac{1 \text{ cu in}}{16.4 \text{ cu cm}} = 0.30 \text{ cu in.}$$

We see that 5 milliliters is 0.30 in³.

e. The relevant conversion factor is 1 kg = 2.2 lb. Therefore, 150 pounds is equal to

$$150 \text{ lb} \times \frac{1 \text{ kg}}{2.2 \text{ lb}} = 68.2 \text{ kg.}$$

We see that 150 pounds is about 68 kilograms.

11. Celsius – Kelvin Conversions. Recall that to convert Celsius temperatures to Kelvin temperatures, we add 273 to the Celsius temperature; to convert Kelvin temperatures to Celsius temperatures, we subtract 273 from the Kelvin temperature

a. The Celsius temperature is 50 K – 273 = –223°C.

c. The Celsius temperature is 500,000 K – 273 = 499,727°C.

e. The Kelvin temperature is 100°C + 273 = 373 K.

f. The Celsius temperature is 320 K – 273 = 47°C.

13. Mountains and Trenches.

a. We must convert 14,494 feet to miles, meters, and kilometers. Using the conversion factor 1 ft = 0.305 m (rounded from Table 3.7), we find that

$$14,494 \text{ ft} = 14,494 \text{ ft} \times \frac{1 \text{ mi}}{5280 \text{ ft}} = 2.75 \text{ miles.}$$

$$14,494 \text{ ft} = 14,494 \text{ ft} \times \frac{0.3048 \text{ m}}{1 \text{ ft}} = 4418 \text{ meters.}$$

$$14,494 \text{ ft} = 4418 \text{ m} \times \frac{0.001 \text{ km}}{1 \text{ m}} = 4.418 \text{ km.}$$

Mt. Whitney is 2.75 miles high, 4421 meters high, and 4.421 km high.

c. The total height of Mauna Kea from its base to its summit is 13,796 ft + 18,200 ft = 31,996 ft. Converting this height to miles, meters and kilometers, using the conversion factor 1 ft = 0.305 m (rounded from Table 3.7), we see that

$$31,996 \text{ ft} = 31,996 \text{ ft} \times \frac{1 \text{ mile}}{5280 \text{ feet}} = 6.06 \text{ miles}$$

$$31,996 \text{ ft} = 31,996 \text{ ft} \times \frac{0.3048 \text{ m}}{1 \text{ foot}} = 9752.4 \text{ m}$$

$$31,996 \text{ ft} = 9752.4 \text{ m} \times \frac{0.001 \text{ km}}{1 \text{ m}} = 9.7524 \text{ km.}$$

The summit of Mt. Everest is about 29,030 feet above sea level. Mauna Kea may be the largest mountain on earth, but Mt. Everest is the highest since elevations are measured from sea level.

15. Foreign Price Conversions.

a. We need the conversion factors 3.785 liter = 1 gallon and 1 franc = $0.21. Then we can do the following conversion:

$$4 \frac{\text{franc}}{\text{liter}} \times \frac{\$0.21}{\text{franc}} \times \frac{3.785 \text{ liter}}{\text{gal}} = \frac{\$3.18}{\text{gal}}.$$

A price of 4 francs/liter is the same as $3.18/gallon.

c. We need the conversion factors 1 kg = 2.2 lb and 1 mark = $0.72. Then we can do the following conversion:

$$2 \frac{\text{mark}}{\text{kg}} \times \frac{\$0.72}{1 \text{ mark}} \times \frac{1 \text{ kg}}{2.2 \text{ lb}} = \frac{\$0.65}{\text{lb}}.$$

A price of 2 mark/kg is the same as $0.65/lb.

17. Carats and Karats. When using Table 3.3, note that there are two different kinds of ounces *and* two different kinds of pounds (avoirdupois and troy).

a. Recall that 24 karat means 100% pure gold. A gold nugget that is 75% pure has 0.75 × 24 = 18 karats.

b. A 14-karat gold chain is 14/24 = 58.3% pure. Therefore, if the chain weighs 15 grams, then it has 15 grams × 0.583 = 8.75 grams of gold.

19. The Cullinan Diamond and the Star of Africa. The conversion factors we need to answer these questions are 1 carat = 0.2 grams = 200 milligrams and 1 lb = 454 grams. The Cullinan diamond weighs

$$3106 \text{ carat} \times \frac{200 \text{ mg}}{\text{carat}} = 621,200 \text{ mg.}$$

The diamond weighs 621,200 milligrams or 621.2 grams which can be converted to pounds:

$$621.2 \text{ gm} \times \frac{1 \text{ lb}}{454 \text{ gm}} = 1.37 \text{ lb.}$$

The Star of Africa diamond weighs

$$530.2 \text{ carat} \times \frac{200 \text{ mg}}{\text{carat}} = 106,040 \text{ mg,}$$

or 106 grams which when converted to pounds gives us

$$106 \text{ gm} \times \frac{1 \text{ lb}}{454 \text{ gm}} = 0.23 \text{ lb.}$$

The Cullinan diamond weighs 1.37 pounds and the Star of Africa diamond weighs about a quarter of a pound.

21. Temperature Formulas. We have a preview of some algebra here! To simplify the notation, let C stand for degrees Celsius and F stand for degrees Fahrenheit. Then we start with the formula

$$C = \frac{F - 32}{1.8}.$$

Multiplying both sides of this expression by 1.8, we have

$$1.8 \times C = 1.8 \times \frac{F - 32}{1.8} = F - 32.$$

Now adding 32 to both sides gives us

$$1.8 \times C + 32 = F - 32 + 32 = F.$$

The formula for converting degrees Celsius to degrees Fahrenheit is $F = (1.8 \times C) + 32$.

Unit 3C The Process of Problem Solving

Overview

Having seen very specific examples of problem solving in the previous two units, we now present some strategies for problem solving in general. *There is no simple and universal formula for solving all problems*! Only continual (and hopefully enjoyable) practice can led one towards mastery in problem solving. This unit is designed to provide you some of that practice.

The unit opens with a very well-known four-step process for approaching problem solving. As you will see, this process is not a magic formula, but rather a set of guidelines. The four basic steps are

- understand the problem,
- devise a strategy,
- carry out the strategy, and
- look back, check, interpret and explain your solution.

The remainder of the unit is a list of eight strategic hints for problem solving. Here are the strategic hints:

- there may be more than one answer.
- there may be more than one strategy.
- use appropriate tools.
- consider simpler, similar problems.
- consider equivalent problems with simpler solutions.
- approximations can be useful.
- try alternative patterns of thought.
- don't spin your wheels.

Most of the unit consists of examples in which the four-step process and these strategic hints are put to use. Study these problems and solutions, and try to see how the techniques might be used in other problems that you might encounter. Most important of all, try to enjoy problem solving!

Key Concepts and Skills

- carry out the four-step problem solving process.
- identify and use the eight strategic problem solving hints.

Unit 3C Solutions

1. Traffic Counters.

a. This solution continues the discussion of Example 1. We can find the other solutions to the traffic counter problem after introducing some short hand notation. Let's agree to let T represent the number of trucks that pass over the counter and C represent the number of cars that pass over the counter. Then as we saw in the example, twice the number of cars plus three times the number of trucks must equal 35, or $(3 \times T) + (2 \times C) = 35$. Some trial and error will discover the remaining solutions:

$T = 11$ and $C = 1$, since $(3 \times 11) + (2 \times 1) = 35$.
$T = 9$ and $C = 4$, since $(3 \times 9) + (2 \times 4) = 35$.
$T = 7$ and $C = 7$, since $(3 \times 7) + (2 \times 7) = 35$.
$T = 5$ and $C = 10$, since $(3 \times 5) + (2 \times 10) = 35$.
$T = 3$ and $C = 13$, since $(3 \times 3) + (2 \times 13) = 35$.
$T = 1$ and $C = 16$, since $(3 \times 1) + (2 \times 16) = 35$.

We see that there are six different solutions to the problem – far from a unique solution! Notice that in each successive solution the number of trucks decreases by 2 while the number of cars increases by 3.

c. In this problem we allow three different types of vehicles. as before we will let C denote 2-axle vehicles, T denote 3-axle vehicles, and now add B for (big) 4-axle vehicles. Now solutions to the problem, with a total of 10 counts, must satisfy the relation

$$(4 \times B) + (3 \times T) + (2 \times C) = 10.$$

Some trial and error lead to the following solutions:
$B = 0$, $T = 0$ and $C = 5$,
        since $(4 \times 0) + (3 \times 0) + (2 \times 5) = 10$.

$B = 1$, $T = 0$ and $C = 3$,
        since $(4 \times 1) + (3 \times 0) + (2 \times 3) = 10$.
$B = 2$, $T = 0$ and $C = 1$,
        since $(4 \times 2) + (3 \times 0) + (2 \times 1) = 10$.
$B = 0$, $T = 2$ and $C = 2$,
        since $(4 \times 0) + (3 \times 2) + (2 \times 2) = 10$.
$B = 1$, $T = 2$ and $C = 0$,
        since $(4 \times 1) + (3 \times 2) + (2 \times 0) = 10$.

We see that there are five different combinations of the three vehicles that satisfy the conditions of the problem.

3. The Cars and Canary Revisited. Reasoning as we did in Example 3, note that the cars approach each other at a combined speed of 180 km/hr. Therefore, the time they travel before they collide (distance divided by speed) is

$$(150 \text{ km}) \div (180 \text{ km/hr}) = 5/6 \text{ hr}.$$

In this amount of time, the canary travels (speed × elapsed time)

$$(120 \text{ km/hr}) \times (5/6 \text{ hour}) = 100 \text{ km}.$$

The canary travels 100 kilometers before the cars collide.

5. The Coiled Wire Revisited. Figure 3.12 of the text is useful in solving this problem. After cutting and flattening the pipe, each resulting right triangle has a base 6 cm long. The length of the original pipe was 20 cm and had 10 wraps of wire on it, so the height of the ten resulting right triangles is 20 cm ÷ 10 = 2 cm. By the Pythagorean theorem, the hypotenuse of each of the ten triangles has a length of

$$\sqrt{(6 \text{ cm})^2 + (2 \text{ cm})^2} = \sqrt{40} \text{ cm} = 6.3 \text{ cm}.$$

The ten hypotenuses correspond to the actual wire; so the total length of the wire is 10×6.3 cm = 63 cm.

9. Hint: What if Reuben were born on December 31?

11. That man is my son.

13. After the first purchase, the woman is $500 in the hole. After the first sale she is $100 ahead. After the second purchase she is $600 in the hole. So after the seconds sale she is $200 ahead. She gains $200 in the end.

15. Select one ball from the first barrel, two balls from the second barrel, three balls from the third, and so forth. If all of the balls weigh one ounce, then the weight of this collection *should be* $1 + 2 + 3 + ... + 9 + 10 = 55$ ounces. Let's say that the weight of the balls you select is 60 ounces. This means that you have five heavy balls which must have come from barrel #5. In general, the excess number of ounces (over 55 ounces) tells you what barrel has the heavy balls.

17. To do it with a delayed start, start both hourglasses together. When the 4-minute hourglass runs out, flip it immediately. When the 7-minute hourglass runs out, there will be 1 minute left in the 4-minute hourglass. Start the 9-minute interval now. Let the 1 remaining minute run out of the 4-minute hourglass and then wait for the 4-minute hourglass to empty two more times. This gives a 9-minute interval. Can you find a way without a delayed start?

19. Divide the coins into three sets of four coins and put two sets on the balance scale. This will determine the set containing the heavy coin (first weighing). Divide this set with the heavy coin into two sets and put them on the balance scale. This determines the set of two coins with the heavy coin (second weighing). Finally, put these two coins on either side of the balance scale to find the heavy coin (third weighing). A more difficult problem is to find the counterfeit coin not knowing whether it is heavy or light.

21. Alma visited at 8:00 p.m., Bess visited at 9:00 a.m., Cleo visited at 10:00 p.m., and Dina visited at 11:00 a.m.

4. NUMBERS IN THE REAL WORLD

Overview

In this chapter we explore the concept of number and then begin to look at how numbers are used in real and relevant problems. The first unit of the chapter is brief and historical, intended to give you an idea of where our current number systems comes from. The next unit deals with percentages and is arguably one of the most practical and important units of the book. As you know, many real world numbers are incredibly large (the federal debt or the storage capacity of a computer disk) or very small (the diameter of a cancer cell or the wavelength of an x-ray); so in the next unit we introduce scientific notation to deal with large and small numbers. The last unit discusses another reality of numbers in the real world : they are often approximate or subject to errors. All in all it's a very practical and useful chapter.

Unit 4A Concepts of Number

Overview

In this brief introductory unit, we introduce the number system that is used today throughout most of the world and talk a little bit about its origins. The main point is that the modern number system is built in layers that we use almost unconsciously. The various sets of numbers that you will meet are

- the natural numbers (or counting numbers) 1, 2, 3, ...
- the integers ...−3, −2, −1, 0, 1, 2, ...
- the rational numbers (or fractions)
- the real numbers (all decimal numbers)
- the complex numbers.

Notice that each set of numbers contains all of the sets that precede it on this list. We also review how the operations of arithmetic work on these sets of numbers.

We also note three different uses of numbers:

- cardinal numbers are used for counting or describing quantities (there are 36 students in class).
- ordinal numbers are used for ordering (she came in third place in the race).

• nominal numbers are used for labeling (follow route 66).

Key Words and Phrases

cardinal numbers	ordinal numbers	nominal numbers
additive system	place-value system	Hindu-Arabic numerals
decimal system	Roman numerals	natural numbers
negative numbers	integers	rational numbers
irrational numbers	real numbers	complex numbers

Key Concepts and Skills

• understand the origins of the modern number system.
• identify the set of numbers to which a given number belongs.
• add, subtract, multiply, and divide numbers in each of the number sets.

Solutions Unit 4A

1. Uses of Numbers.

a. *Three* is used as a <u>cardinal</u> number because it is used for counting.

b. The statement does not mean that there are 45 people in the line. Therefore, 45 is used as an <u>ordinal</u> number since it used for ordering.

c. The price of the car is a <u>cardinal</u> number because it is used to indicate an amount.

d. The number 66 is used as a <u>nominal</u> number because it is a label.

3. Classifying Numbers.

a. We can interpret 2.3 as 2 + 3/10 which can be written as the fraction 23/10. Thus, 2.3 is a rational number. Recall also that rational numbers have either terminating or repeating decimal expansions which also qualifies 2.3 as a rational number.

b. The number −3/2 is a (negative) rational number because it is expressed as a fraction.

c. The number 3 is a natural number or a counting number.

d. The number −5 is an integer.

e. A number with either a terminating or repeating decimal expansion is a rational number. Therefore,

100.1 is a rational number. This number can also be expressed as the fraction 1001/10.

f. The number −6.1 is a rational number because it has a terminating decimal expansion and can be written as the fraction −61/10.

g. The fraction 5/3 is a rational number. It has a repeating decimal expansion 1.6666666....

h. The number π is irrational; it is a real number, but not a rational number. It cannot be expressed as a fraction and it has a non-terminating, non-repeating decimal expansion.

5. How Operations Affect Numbers.

a. A product of any number of integers is always an integer. The product of two negative numbers is positive, so the product of three negative numbers is positive × negative which is negative. Therefore, the product of three negative integers is a negative integer. For example,

$$(-2) \times (-3) \times (-4) = -24.$$

b. Dividing an integer by a rational number (fraction) will in general produce a rational number. Dividing any positive number by any negative number produces a negative number. Therefore, dividing a positive integer by a negative rational number will result in a negative rational number. For example,

$$4 \div (-3/2) = -8/3.$$

c. The product of a positive rational number and a positive irrational number is positive and irrational. For example, $(2/3) \times \pi = (2 \times \pi)/3$, which is positive and irrational.

d. In general, adding two irrational numbers produces an irrational number. For example, $\pi + \pi = 2\pi$, which is irrational. However, it is possible that the sum of two irrational numbers could be rational. For example, $\pi + (5 - \pi) = 5$.

7. Fractions to Decimals.

a. The fraction 3/4 is equivalent to 0.75. The decimal terminates.

b. The fraction 4/11 can be divided out on a calculator; it is equivalent to 0.363636... The repeating pattern is 36.

c. The fraction 5/8 is equivalent to 0.625. The decimal terminates.

d. The fraction 1/6 can be divided out on a calculator; it is equivalent to 0.16666... The repeating pattern is 6.

e. The fraction 4/7 can be divided out on a calculator; it is equivalent to 0.571428571428... The repeating pattern is 571428.

f. The fraction 11/20 is equivalent to 0.55. The decimal terminates.

g. The fraction 37/60 can be divided out on a calculator; it is equivalent to 0.616666... The repeating pattern is 6.

h. The fraction 23/37 can be divided out on a calculator; it is equivalent to 0.621621621... The repeating pattern is 621.

Thinking About Numbers.

9. The number of students in your class can be described using a natural (counting) number. Given a small class size and accurate counting, the number should be exact.

11. The temperature is given using a rational number, probably in the form of a decimal. The number will be approximate, rounded off to, say, the nearest degree.

13. The fraction of women is a rational number expressed either as a fraction (say 3/5) or as a percentage (say 60%). It could be exact, up to counting errors.

A Language of Symbols.

15. This statement can be expressed as $10/x$ where $x \neq 0$.

17. This statement can be expressed as $a - 12$, where a is any number.

Mathematical Patterns.

19. The pattern is *adding* a set of consecutive integers. The pattern stops after six terms. The full statement is $1 + 2 + 3 + 4 + 5 + 6$.

21. The pattern is adding numbers that increase by a factor of 10 each time. The pattern ends and the entire statement reads: $1 + 10 + 100 + 1,000 + 10,000 + 100,000 + 1,000,000$.

23. The pattern is adding successive powers of 2. The pattern ends and the entire statement reads: $2^0 + 2^1 + 2^2 + 2^3 + 2^4 + 2^5 + 2^6 + 2^7 + 2^8 + 2^9 + 2^{10} = 1 + 2 + 4 + 8 + 16 + 32 + 64 + 128 + 256 + 512 + 1024$.

25. The pattern is listing the powers of two. The pattern ends and the entire statement reads: 2, 4, 8, 16, 32, 64, 128, 256, 512, 1024.

27. Look at the difference between successive terms. The pattern is adding numbers that increase as the square of successive integers. The pattern continues forever and with three more terms reads: $1 + 5 + 14 + 30 + 55 + 91 + 140 + 204, \ldots.$

29. Reading and Writing Roman Numerals.

a. 53	b. 66	c. 121
d. 181	e. 1443	f. 44
g. IV	h. XXXVII	i. XLI
j. XLIX	k. CVI	l. CCCXXXIV

31. Roman Numerals in Use.

a. The Roman numeral that appears on the back of a one-dollar bill is MDCCLXXVI which translates to 1776, the year in which the Declaration of Independence was signed.

b. The Roman numeral MCCCLXXIX can be translated to 1379, most likely the year in which the corner stone of the building was laid. This date would not be found on a building in the United States because, although the land was inhabited at that time, the same calendar was not in use.

Unit 4B Uses and Abuses of Percentages

Overview

If you read a news article, a financial statement, or an economic report, you will see that one of the most common ways to communicate quantitative information is with percentages. In this units we will explore the many ways that percentages are used — and abused. Percentages are used for three basic purposes:

- as fractions (for example, 45% of the voters favored the incumbent),
- to describe change (for example, taxes increased by 10%), and
- for comparison (for example, women live 3% longer than men).

In this unit, we examine each of these uses of percentages in considerable detail with plenty of examples taken from the news and from real situations. The use of percentages as fractions is probably familiar. The use of percentages to describe change relies on the notions of **absolute change** and **relative** (or **percentage change**). When some quantity changes from a previous value to a new value, we can define its changes in either of two ways:

$$\text{absolute change} = \text{new value} - \text{previous value}$$

$$\text{relative change} = \frac{\text{absolute change}}{\text{previous value}} = \frac{\text{new value} - \text{previous value}}{\text{previous value}}.$$

The use of percentages for comparison relies on the notions of **absolute difference** and **relative** (or **percentage difference**). To make comparisons between two quantities, we must identify the **compared quantity** (the quantity that we are comparing) and the **reference quantity** (the quantity that we are comparing to). For example, if we ask how much larger is one meter than one yard, *one meter* is the compared quantity and *one yard* is the reference quantity. Then we can calculate two kinds of difference:

$$\textbf{absolute difference} = \text{compared quantity} - \text{reference quantity}$$

$$\textbf{relative difference} = \frac{\text{absolute difference}}{\text{reference quantity}} = \frac{\text{compared quantity} - \text{reference quantity}}{\text{reference quantity}}$$

A powerful rule for interpreting statements involving percentages is what we call the **Of Versus More Than Rule**. It says that if the compared quantity is $P\%$ *more than* the reference quantity, then it is $(100 + P)\%$ *of* the reference quantity. For example, if Bess' salary is 30% more than Bob's salary, then Bess' salary is 130% of Bob's salary. Similarly, if Jill's height is 30% *less than* Jack's height, then Jill's height is 70% of Jack's height.

All of these ideas are assembled in this unit to solve a variety of practical problems. Our experience in teaching this subject is that the greatest difficulty is understanding the problem and translating it into mathematical terms. It is important to read the problem carefully, draw a picture if necessary, decide how percentages are used in the problem (as a fraction, for change, or for comparison), and to write a mathematical sentence that describes the situation. It is important to study the examples in the unit and work plenty of practice problems.

The unit closes with several examples of ways that percentages are abused. Be sure you can identify problems in which the previous value shifts and problems in which the quantity of interest is itself a percentage.

Key Words and Phrases

absolute change	relative change	percentage change
reference quantity	compared quantity	absolute difference
relative difference	percentage difference	of versus more than rule

Key Concepts and Skills

- identity and solve problems that use percentages as fractions.
- identity and solve problems that use percentages to describe change.
- identity and solve problems that use percentages for comparison.
- apply the *of versus more than rule* to practical problems.
- identify fallacies in statements involving percentages.

Solutions Unit 4B

1. Percentages as Fractions.

a. The fraction of women is $25/113 = 0.221 = 22.1\%$.

b. The fraction of blooming tulips is $345/398 = 0.867 = 86.7\%$.

c. The fraction of people who voted for the losing candidate was $1234/3009 = 0.410 = 41.0\%$.

3. Working with Percentages as Fractions.

a. The total is $23/0.03 = 766.67$. You should check to see that this answer makes sense. If it is correct, it must follow that 3% of 766.67 is really 23. Check that $0.03 \times 766.67 = 23$, as required.

b. The total is $100/0.15 = 666.67$. You should check to see that this answer makes sense. If it is correct, it must follow that 15% of 666.67 is really 100. Check that $0.15 \times 666.67 = 100$, as required.

c. The total is $150/0.89 = 168.54$. You should check to see that this answer makes sense. If it is correct, it

must follow that 89% of 168.54 is really 150. Check that $0.89 \times 168.54 = 150$, as required.

5. Percentages for Change.

a. A change from 50 to 55 is an increase with an absolute change of $55 - 50 = 5$. The relative change as a fraction, a decimal, and a percentage is

Relative change $= \dfrac{\text{Absolute change}}{\text{Previous value}} = \dfrac{5}{50} = 0.1 = 10\%.$

The change is 10% of the previous value.

b. A change from 0.1 to 0.5 is an increase with an absolute change of 0.5 − 0.1 = 0.4. The relative change as a fraction, a decimal, and a percentage is

Relative change $= \dfrac{\text{Absolute change}}{\text{Previous value}} = \dfrac{0.4}{0.1} = 4.0 = 400\%.$

The change is 400% of the previous value.

c. A change from 1000 to 1001 is an increase with an absolute change of 1001 − 1000 = 1. The relative change as a fraction, a decimal, and a percentage is

Relative change $= \dfrac{\text{Absolute change}}{\text{Previous value}} = \dfrac{1}{1000} = 0.001 = 0.1\%.$

The change is 0.1% of the previous value.

7. Change Everywhere.

a. We are told that the U.S. population increased from 220 million (previous value) to 265 million (new value). The absolute change was 265 million − 220 million or 45 million. The percentage change was

$\dfrac{\text{Absolute change}}{\text{Previous value}} \times 100\% = \dfrac{45\text{ million}}{220\text{ million}} \times 100\% = 20.45\%.$

Between 1974 and 1995 the U.S. population increased by 20.45%.

b. We are told that the number of deaths due to heart disease decreased from 250,000 (previous value) to 140,000 (new value). The absolute change was 140,000 − 250,000 = −110,000. The percentage change was

$\dfrac{\text{Absolute change}}{\text{Previous value}} \times 100\% = \dfrac{-110,000}{250,000} \times 100\% = -44\%.$

Between 1970 and 1993 the number of deaths due to heart disease decreased by 44%.

c. We are told that defense spending decreased from $266 billion (previous value) to $254 billion (new value). The absolute change was $254 billion − $266 billion = −$12 billion. The percentage change was

$\dfrac{\text{Absolute change}}{\text{Previous value}} = \dfrac{-\$12\text{ billion}}{\$266\text{ billion}} = -0.045 = -4.5\%.$

Between 1995 and 1997 the U.S. defense budget changed by −4.5% which is a decrease of 4.5%.

9. Fifty Years of World Population Growth.

a. The relative change in world population between 1950 and 1970 is

$\dfrac{\text{new value - old value}}{\text{old value}} = \dfrac{3.7 - 2.0}{2.0} = 0.85 = 85\%.$

b. The relative change in world population between 1950 and 2000 is

$\dfrac{\text{new value - old value}}{\text{old value}} = \dfrac{6.1 - 2.0}{2.0} = 2.05 = 205\%.$

c. The relative change in world population between 1970 and 2000 is

$\dfrac{\text{new value - old value}}{\text{old value}} = \dfrac{6.1 - 3.7}{3.7} = 0.65 = 65\%.$

d. The relative change in world population between 1990 and 2000 is

$\dfrac{\text{new value - old value}}{\text{old value}} = \dfrac{6.1 - 5.3}{5.3} = 0.15 = 15\%.$

11. Percentages for Comparison.

a. The words "than $1" tell us that the dollar is the reference quantity. Note that the conversion factor 0.6 pounds/$1 also means $1/0.6 pounds, or $1.67/pound. The absolute difference is

1 pound − $1 = $1.66 − $1 = $0.66.

To find the relative and percentage difference, we divide the absolute difference by the reference quantity:

% difference $= \dfrac{\text{abs difference}}{\text{ref quantity}} = \dfrac{\$0.66}{\$1} = 0.66 = 66\%.$

One English pound is 66% larger than $1.

b. The words "than a mile" tell us that the mile is the reference quantity. Notice also that if 1 mi = 1.6 km, then 1 km = 1/1.6 mi = 0.63 mi. The absolute difference is

1 km − 1 mi = 0.63 mi − 1 mi = −0.37 mi.

To find the relative and percentage difference, we divide the absolute difference by the reference quantity:

% difference $= \dfrac{\text{abs difference}}{\text{ref quantity}} = \dfrac{-0.37\text{ mi}}{1\text{ mi}} = -0.37 = -37\%.$

A kilometer is 37% *smaller* (because of the minus sign) than a mile.

c. The words "than California" tell us that the population of California is the reference quantity. The absolute difference is

18.1 million − 31.6 million = −13.5 million.

To find the relative and percentage difference, we divide the absolute difference by the reference quantity:

% difference $= \dfrac{\text{abs difference}}{\text{ref quantity}} = \dfrac{-13.5\text{ million}}{31.6\text{ million}} = -0.427 = -42.7\%.$

New York is 42.7% *less* populous (because of the minus sign) than California.

13. Using the *Of* versus *More Than* Rule. Recall that if the compared quantity is *P*% *more than* the reference quantity, then it is (100 + *P*)% *of* the reference quantity.

a. Because Brian earns 300% *more than* Wilson, Brian's income is (100 + 300)% = 400% *of* Wilson's income. This means Brian earns four times Wilson's income.

b. Because Kathy's income is 0.8 =80% *of* Martha's income, Kathy earns (100 − 80)% = 20% *less than* Martha

c. Because the retail price is 40% *more than* the wholesale price, the retail price is (100 + 40)% = 140% *of* the wholesale price. This means the retail price is 1.4 times the wholesale price.

d. The sale price is 30% *less than* the regular price. Therefore, the sale price is (100 − 30)% = 70% *of* the regular price. This means that

$32.40 = Sale price = 70% × regular price.

To find the regular price, we can divide both sides by 70% = 0.7:

$$\text{regular price} = \frac{\text{sale price}}{0.7} = \frac{\$32.40}{0.7} = \$46.29.$$

Check that indeed, $32.40 is 30% less than $46.29.

15. Percent Larger and Smaller. Suppose A = 5 and B = 4. Then A is 25% larger than B as the problem stipulates (since 4 + (25% × 4) = 5). But comparing B to A, we see that (B − A)/B = (4 − 5)/5 = −20%. This means that B is 20% smaller than A. This argument works for any values of A and B. In general, if A is p% larger than B, then B cannot be p% smaller than A.

17. Sales Tax. The total pre-tax price of the items is $495 + $429 = $924. Adding 0.083 × $924 = $76.69 for tax brings the total bill to $1000.69.

19. Comparison Shopping. If $5 is 5% of the purchase price in Niwot, then the purchase price must be $5/0.05 = $100. (Check that 5% of $100 is $5.) If $5 is 7% of the purchase price in Longmont, then the purchase price is $5/0.07 = $71.43. (Check that 7% of $71.43 is $5.) The price is *lower* in Longmont because $5 is a *larger* percentage of the purchase price in Longmont.

21. Profitable Company? To be specific, assume that the company's last year's profits were $1000 (you can choose any number). After a 25% loss, the profits the following year would be 75% of $1000 or 0.75 × $1000 = $750. After a 45% gain, the profits the following year would be $750 + (0.45 × $750) = $1087.50. Indeed, the company's profits have improved over two years, but only by 8.75% and not by 45% − 25% = 20% as the president may have suggested.

23. Shifting Percentages.

a. If $10 is 1% of your balance, then the balance must be $10/0.01 = $1000. (Check that 1% of $1000 is $10).

b. If $10 is 4% of your balance, then the balance must be $10/0.04 = $250. (Check that 4% of $250 is $10).

c. The balance in the account decreased from $1000 to $250 which is a change of ($250 - $1000)/$1000 = −75%.

25. Analyzing Percentage Statements.

a. This statement would be true were it not for an arithmetic error: 40% × 20% = 8%, not 80%.

b. This statement is false because some of the hotels with a restaurant may also have a pool. The statement would be true only if the hotels with restaurants are all different from the hotels with pools.

c. This statement is true. A total of 50% of the students either drive or take a bus. So the remaining 50% must neither drive nor take a bus.

27. Adding Percentages. The fallacy in this reasoning is that the two groups, Union members and Democrats, most likely overlap; that is, there are people who are Democrat Union members. Suppose that the population of the district were distributed as follows:

	Democrat	Non-Democrat
Union	30%	5%
Non-Union	10%	55%

Notice that 40% of the population is Democrat and 35% of the population is Union members, as stipulated in the problem. But notice also that if the candidate captures all of the Democrat votes and all of the Union votes, s/he has less than a majority (45%) of the vote.

29. Opening Quote.

a. The quantity that changes is the percentage of 12th graders who smoke. It increased from an unknown previous value in 1991 to 22%, which we are told is an increase of 20%. This means that the new smoking rate is 120% = 1.2 of the previous smoking rate. We can write this as

22% = new rate = 1.2 × previous rate.

To find the 1991 smoking rate, we divide both sides of this equation by 1.20:

$$\text{Previous smoking rate} = \frac{22\%}{1.2} = 18.3\%.$$

The smoking rate increased from 18.3% to 22%. Check that this really is a 20% increase!

b. The quantity that changes is the percentage of 10th graders who smoke. It increased from an unknown previous value in 1991 to 18.3%, which we are told is an increase of 45%. This means that the new smoking rate is 145% = 1.45 of the previous smoking rate. We can write this as

18.3% = new rate = 1.45 × previous rate.

To find the 1991 smoking rate, we divide both sides of this equation by 1.20:

Previous smoking rate $= \dfrac{18.3\%}{1.45} = 12.6\%$.

The smoking rate increased from 12.6% to 18.3%. Check that this really is a 45% increase!

31. Percentages in the News.

a. This is a forward percentage problem. The amount of U.S. trade delivered by truck is 75% × $100 billion = $75 billion.

b. If the 1994 GNP was 70% lower than the 1958 GNP, then the 1994 GNP is 100% − 70% = 30% of the 1958 GNP. We can write this as

$125 = 1994 GNP = 0.30 × 1958 GNP.

To solve for 1958 GNP, we must divide both sides by 0.30:

$$1958 \text{ GNP} = \dfrac{\$125}{0.30} = \$417.$$

Check that indeed, $125 is 70% lower than, or 30% of $417.

c. If the number of deaths decreased by 12%, then the number of deaths in 1996 is 100% − 12% = 88% of the number of deaths in 1995. We can write

22,000 = 1996 deaths = 0.88 × 1995 deaths.

To solve for 1995 deaths, we must divide both sides by 0.88:

$$1995 \text{ deaths} = \dfrac{22,000}{0.88} = 25,000.$$

Check that indeed, 22,000 is 12% lower than, or 88% of 25,000.

d. Because the number of Americans studying in Britain increased by 16%, the number of Americans in 1995 is 100% + 16% = 116% of the number of Americans in 1994. We have

19,410 = Americans in 1995 = 1.16 × Americans in 1994.

To solve for Americans in 1994, we must divide both sides by 1.16:

$$\text{Americans in 1994} = \dfrac{19,410}{1.16} = 16,733.$$

Check that indeed, 16,733 is 16% higher than, or 116% of 19,410.

e. Because bankruptcies increased by 6% from 1994 to 1995, the 1995 number is 100% + 6% = 106% of the 1994 number. We can write

832,000 = 1995 bankruptcies = 1.06 × 1994 bankruptcies.

To solve for 1994 bankruptcies, we must divide both sides by 1.06:

$$\text{Bankruptcies in 1994} = \dfrac{832,000}{1.06} = 784,906.$$

Check that indeed, 832,000 is 6% higher than, or 106% of 784,906.

Unit 4C Putting Numbers in Perspective

Overview

Numbers in the world around us are often very large or very small. In order to write large and small numbers compactly, without a lot of zeros, we use scientific notation. This is the main new mathematical idea in this unit. If you haven't used scientific notation before, you will want to practice writing numbers in scientific notation, and multiplying and dividing numbers in scientific notation.

We next spend some time showing how to make rough calculations using estimation. Often we can use approximate values of quantities and arrive at useful **order of magnitude** answers.

All of us suffer from number numbness: large and small numbers eventually lose all meaning. For example, most of us have no sense of how large $5 trillion (the federal debt) or 6 billion (the world's population) really are. Therefore, just as important as *writing* large and small numbers is *visualizing* large and small numbers. We look at some fascinating methods, such as **scaling**, for giving large and small numbers meaning. The goal is often to associate numbers with a striking visual image. For example, if the

Earth is the size of a ball-point pen tip, the Sun would be a grapefruit 15 meters away.

Key Words and Phrases

scientific notation	order of magnitude	scaling
scale factor	light-year	

Key Concepts and Skills

- write numbers in scientific notation.
- multiply and divide numbers in scientific notation.
- use estimates to compute order of magnitude answers.
- use scaling methods for visualizing large and small numbers.

Solutions Unit 4C

3. Reading Powers of 10.

a. $5 \times 10^6 = 5,000,000 = 5$ million.

b. $7 \times 10^9 = 7,000,000,000 = 7$ billion.

c. $-2 \times 10^{-2} = -0.02 = -2$ hundredths.

d. $8 \times 10^{11} = 800,000,000,000 = 800$ billion.

e. $1 \times 10^{-7} = 0.0000001 = 1$ ten millionth .

f. $9 \times 10^{-4} = 0.0009 = 9$ ten thousandths

5. Writing Powers of 10. The table gives the numbers written in scientific notation. (Recall that "officially" the number multiplying the power of ten must be between 1 and 10.)

a. 6×10^2	b. 9×10^{-1}	c. 5×10^4
d. 3×10^{-3}	e. 5×10^{-4}	f. 7×10^{10}

7. Converting to Scientific Notation. The table shows each number in scientific notation.

a. 1×10^6	b. 1.5×10^5	c. $4.5 \times 10^0 = 4.5$
d. 1.8×10^{-1}	e. 5.40×10^8	f. 5.30×10^{25}

9. Converting from Scientific Notation.

a. $2.2 \times 10^{-4} = 0.00022$

b. $2 \times 10^{-1} = 0.2$

c. $9.828 \times 10^7 = 98,280,000$

d. $6.667 \times 10^1 = 66.67$

e. $3.5 \times 10^4 = 35,000$

f. $1.501 \times 10^{-10} = 0.0000000001501$

11. Large and Small Numbers Everywhere.

a. The U.S. national debt in 1997 was about $\$5.3 \times 10^{12}$.

b. Corporate profits in the United States in 1996 were $\$6.32 \times 10^{11}$.

c. Consumer debt (loans and credit cards) in the United States in 1994 was $\$8 \times 10^{11}$.

d. The hard drive on my computer has a capacity of 5.4×10^9 bytes.

13. Approximation with Powers of 10.

a. $10^{26} + 10^7 = 100,000,000,000,000,000,000,000,000 + 10,000,000 = 100,000,000,000,000,000,010,000,000$. Since 10^{26} is so much larger than 10^7 (10^{19} times larger), the 10^7 may be considered insignificant as shown in the calculation above. To a high degree of accuracy, $10^{26} + 10^7$ is equal to 10^{26}.

b. The number 10^{81} is 10^{19} times larger than 10^{62}. So when we subtract 10^{62} from 10^{81}, the change in 10^{81} is insignificant. To a high degree of accuracy, $10^{81} - 10^{62}$ is equal to 10^{81}.

15. Practice with Scientific Notation.

a. $(9 \times 10^8) + (5 \times 10^9) = (0.9 \times 10^9) + (5 \times 10^9) = 5.9 \times 10^9$.

b. $(8.1 \times 10^{30}) \times (9 \times 10^{15}) = (8.1 \times 9) \times (10^{30} \times 10^{15}) = 72.9 \times 10^{45} = 7.29 \times 10^{44}$.

c. $(2.5 \times 10^{-4}) \times (3 \times 10^{-4}) = (2.5 \times 3) \times (10^{-4} \times 10^{-4}) = 7.5 \times 10^{-8}$.

d. $(8.1 \times 10^{14}) \div (3 \times 10^{-7}) = (8.1 \div 3) \times (10^{14} \div 10^{-7}) = 2.7 \times 10^{14 - (-7)} = 2.7 \times 10^{21}$.

e. $(6.6 \times 10^{-2}) \div (4.4 \times 10^{-3}) = (6.6 \div 4.4) \times (10^{-2} \div 10^{-3}) = 1.5 \times 10^{-2 - (-3)} = 1.5 \times 10^{1} = 15$.

f. We must express both numbers with the same power of ten before subtracting. We then have $(2.1 \times 10^4) - (1.5 \times 10^5) = (0.21 \times 10^5) - (1.5 \times 10^5) = -1.29 \times 10^5$.

g. $(2.4 \times 10^2) \div (8.1 \times 10^8) = (2.4 \div 8.1) \times (10^2 \div 10^8) = 0.30 \times 10^{-6} = 3.0 \times 10^{-7}$.

h. $(9.8 \times 10^2) \times (2.0 \times 10^{-5}) = (9.8 \times 2.0) \times 10^{2-5} = 19.6 \times 10^{-3} = 1.96 \times 10^{-2}$.

17. Approximation with Scientific Notation.

a. We can approximate this quotient by $(6 \times 10^9) \div (2 \times 10^2) = 3 \times 10^7$. The exact value is 2.8×10^7 which is 6.7% below the estimate.

b. We can approximate this quotient by $(4 \times 10^{12}) \div (2 \times 10^8) = 2 \times 10^4 = 20,000$. The exact value is 15,385 which is considerably less than the estimate. The large error occurred because 260 million was not rounded very accurately.

c. We can approximate this product by $(9 \times 10^3) \times (5.5 \times 10^4)$ which is approximately $50 \times 10^7 = 5 \times 10^8$. The exact value is 4.95×10^8 which is 1% below the estimate.

d. We can approximate this quotient by $(3 \times 10^9) \div (3 \times 10^4) = 10^5 = 100,000$. The exact value is 1.2×10^5 which is considerably more than the estimate because 25,000 was not rounded very accurately.

e. We might replace $5,987 \times 341$ by $(6 \times 10^3) \times (3.5 \times 10^2) = 21 \times 10^5 = 2.1 \times 10^6$. The exact value, 2.04×10^6 is within 3% of the estimate.

f. The quotient $43 \div 765$ is approximately $40 \div 800 = 0.05$. The exact value is 0.056 which is within about 12% of the estimate.

19. They Don't Look Very Different! The easiest way to compare the size of two number is to divide the larger by the smaller. This gives the factor by which one number is larger than the other.

a. Note that 5 billion $= 5 \times 10^9$ and 250 million $= 2.5 \times 10^8$. Dividing the larger number by the smaller number, we have that $(5 \times 10^9) \div (2.5 \times 10^8) = 2 \times 10 = 20$. This means that 5 billion is 20 times larger than 250 million.

b. Dividing the larger number by the smaller number, we see that $(9.3 \times 10^2) \div (3.1 \times 10^{-2}) = 3 \times 10^{2 - (-2)} = 3 \times 10^4$. This means that 9.3×10^2 is 30,000 times larger than 3.1×10^{-2}.

c. Note that 10^{-8} is larger than 2×10^{-13}. Dividing the larger number by the smaller number, we have that

$10^{-8} \div (2 \times 10^{-13}) = 0.5 \times 10^{-8 - (-13)} = 0.5 \times 10^5 = 5 \times 10^4$. This means that 10^{-8} is 50,000 times larger than 2×10^{-13}.

d. Note that 3.5×10^{-2} is larger than 7×10^{-8}. Dividing the larger number by the smaller number gives $(3.5 \times 10^{-2}) \div (7 \times 10^{-8}) = 0.5 \times 10^{-2 - (-8)} = 0.5 \times 10^6 = 5 \times 10^5$. This means that 3.5×10^{-2} is 500,000 times larger than 7×10^{-8}.

e. Note that $1000 = 10^3$ is larger than $0.001 = 10^{-3}$. Dividing the larger number by the smaller number, we have that $10^3 \div 10^{-3} = 10^{3 - (-3)} = 10^6$. This means that 1000 is 1 million times larger than 0.001.

f. Note that 10^{12} is larger than 10^{-9}. Dividing the larger number by the smaller number, we have that $10^{12} \div 10^{-9} = 10^{12 - (-9)} = 10^{21}$. This means that 10^{12} is 10^{21} times larger than 10^{-9}.

P 23. Comparisons Through Estimation.

a. A common estimate for the height of a story in a building is between 10 and 12 feet. Therefore, a 10-story building is between 100 and 120 feet high. A football field is 100 yards or 300 feet long — almost three times longer than the building is high.

b. Let's assume that the walk across the United States is 3200 miles long, that the walker can cover 2 miles per hour and can walk 8 hours per day. The trip will take

$$3200 \text{ miles} \div 2 \ \frac{\text{mile}}{\text{hour}} = 1600 \text{ hours}$$

which amounts to 200 days of walking. We see that the walk can be done in a year even if some of the assumptions are relaxed.

25. Scale Factors.

a. The scale factor can be written as 1 cm = 1 km. Since 1 km $= 10^5$ cm = 100,000 cm, the scale factor can also be written 1 to 100,000 or 1/100,000.

b. We must first convert miles to inches. We have that 1 mile is the same as

$$1 \text{ mile} \times 5280 \ \frac{\text{ft}}{\text{mile}} \times 12 \ \frac{\text{in}}{\text{ft}} = 63,360 \text{ in.}$$

Therefore 0.5 miles is 31,680 in. The scale factor can be written 2 inches = 0.5 miles or 2 inches = 31,680 inches or 2 to 31,680 or 1 to 15,840 or 1/15840.

c. The scale 5 cm = 100 km is the same as 1 cm = 20 km (dividing both lengths by 5). Since 1 km $= 10^5$ cm = 100,000 cm, it follows that 20 km = 2,000,000 cm. The scale can be written 1 to 2,000,000 or 1/2,000,000.

d. Recall that 3.28 ft = 1 m. Therefore, 100 m = 328 ft. The scale 1 ft = 100 m can be written 1 to 328 or 1/328.

27. Atoms in a Sugar Cube. The number of atoms that would fit in a sugar cube is about

$$(100 \text{ million})^3 = (10^8)^3 = 10^{24},$$

which is 10^2, or 100, times more than the number of stars in the universe.

29. Universal Timeline. One billion years is 1/15 of the age of the universe, so it is represented by $(1/15) \times 100$ m = 6.7 m along the timeline. Human history is $10^4/(15 \times 10^9) = 6.7 \times 10^{-7}$ of the age of the universe, so it is represented by $(6.7 \times 10^{-7}) \times 100$ m = 0.000067 m = 0.067 mm along the timeline.

31. Printing Money. We can let units guide the calculation and find that a rate of 5 trillion bills per year can be converted to

$$5 \times 10^{12} \frac{\text{bills}}{\text{year}} \times \frac{1 \text{ year}}{365 \text{ days}} \times \frac{1 \text{ day}}{24 \text{ hr}} \times \frac{1 \text{ hr}}{3600 \text{ sec}} = 1.6 \times 10^5 \frac{\text{bills}}{\text{sec}}.$$

We see that in order to print enough $1 bills to retire the federal debt in a year, bills would have to be printed at a rate of about 160,000 bills per second. If $100 bills were printed instead of $1 bills the rate could be decreased by a factor of 100 (since a $100 bill has the effect of paying 100 $1 bills). Thus, $100 bills would need to be printed at a rate of 1600 bills per second.

Until the 1970's all money had to be backed up by gold reserves (called the gold standard) so money could not be printed at will. More recently, the primary consequence of printing large sums of money (for example, to pay off the federal debt) is that it puts additional money in circulation which raises prices, then wages, and creates a spiral of inflation.

33. Paving with Dollar Bills. A $1 bill has the dimensions 15.5 cm $\times$ 6.5 cm which gives an area of approximately 100 cm^2.

a. If the federal debt of 5×10^{12} were laid out in $1 bills like paving stones, the area that could be covered is $5 \times 10^{12} \times 100$ cm^2/$ = 5 \times 10^{14}$ cm^2. How much area is

this? Recall that since 1 km = 10^5 cm, 1 km^2 = $(10^5$ cm$)^2$ = 10^{10} cm^2. Doing the conversion from square centimeters to square kilometers, we find the area that can be paved by the debt is

$$5 \times 10^{14} \text{ cm}^2 \times \frac{1 \text{ km}^2}{10^{10} \text{ cm}^2} = 5 \times 10^4 \text{ km}^2.$$

The federal debt could pave 50,000 square km in $1 bills.

b. The area of the United States is about 10^7 km^2 $\div$ $(5 \times 10^4$ km^2) = 200 times greater than the area than can be paved by the federal debt.

35. Zipper Money. Let's assume that 260 million people each buy 10 zippers and that each zipper produces $0.01 (1 cent) royalty for the inventor. The inventor will earn

$$2.6 \times 10^8 \text{ people} \times 10 \frac{\text{zippers}}{\text{person}} \times 0.01 \frac{\$}{\text{zipper}} = \$2.6 \times 10^7.$$

We see that this prosperous inventor will earn $26 million per year in royalties.

37. Water Use.

a. Using a population figure of 260 million people, the daily per capita water use can be found by dividing by the population:

$$3.40 \times 10^{11} \frac{\text{gal}}{\text{day}} \div 2.60 \times 10^8 \text{people} = 1.3 \times 10^3 \frac{\text{gal}}{\text{person - day}}.$$

b. Since public water comprises 10% of the total water use, the daily per capita use of public water is 10% of the daily per capita use of total water (found in the previous problem) which is $0.10 \times 1.3 \times 10^3$ gallons per person per day = 1.3×10^2 gallons per person per day. On average, each person uses 130 gallons of water per day!

Unit 4D Dealing With Uncertainty

Overview

While they may appear totally reliable, the numbers we see in the news or in reports often are quite uncertain and prone to errors. This unit reminds us of this fact in many different ways, as we consider the sources of errors and the measurement of uncertainty in numbers.

The first observation in the unit concerns the difference between **accuracy** and **precision**. A pharmacist's scale that can weigh fractions of an ounce has much more *precision* that a butcher's scale that can measure only in pounds. Thus precision refers to how precisely a quantity can be measured or how precisely a number is reported.

On the other hand the *accuracy* of a measurement refers to how close it is to the exact value (which we often don't know). Suppose a marble weighs 4.5 ounces. If one scale reports a weight of 4.4 ounces and another scale reports a weight of 4.7 ounces, the first measurement is more accurate; both measurements have the same precision — to the nearest tenth of an ounce.

The precision of a number can be described by the number of **significant digits** it has. Significant digits are those digits in a number that we can assume to be reliable, although, this is where care must be used when reading numbers. It's usually easy to determine the number of significant digits in a number; there are a few subtle cases that are listed in the summary box in the unit.

A few technicalities arise when it comes to doing arithmetic with approximate numbers. There are two rules that tell us how much precision should be assigned to the sum, difference, product, or quotient of two approximate numbers.

• when adding or subtracting two approximate numbers, the result should be rounded to the same precision as the *least precise* number in the problem.

• when multiplying or dividing two approximate numbers, the result should be rounded to the same number of significant digits as the number in the problem with the *fewest significant digits*.

The unit closes with three case studies about the census, the consumer price index, and the species extinction rate. These case studies illustrate the implications of working with approximate numbers.

Key Words and Phrases

accuracy	precision	significant digits
participation bias	random errors	systematic errors
consumer price index		

Key Concepts and Skills

• understand the difference between accuracy and precision.

• determine the number of significant digits in a given number.

• determine the precision of the sum or difference of two approximate numbers.

• determine the precision of the product or quotient of two approximate numbers.

• analyze a news report to determine sources of error and reliability of stated numbers.

Solutions Unit 4D

5. Sources of Error.

a. Notice that there are three variables in this statement: amount of memory, price, and time. There are sampling errors involved in this statement since chips

with different memory sizes must be sampled over several years for their prices.

b. This needn't be a statistical statement! If school records are accurate, it should not contain any errors.

c. This statement might have resulted from exit polls. If so, it will contain errors due to sampling. On the other hand, it may result from an accurate count of votes, in which case the only sources of error are in the counting and the rounding that led to the figure 2/3.

d. This estimate of the literacy rate in a large country has errors due to sampling and misreporting. We assume it means adult literacy, but even then, we might question how literacy is defined.

7. More Rounding. The numbers below are rounded to the nearest thousandth, tenth, ten, and hundred in that order.

a. 2365.98521 rounds to 2365.985, 2366.0, 2370, and 2400.

b. 322354.09005 rounds to 322354.090, 322354.1, 322350, 322,400

c. 6000 rounds to 6000.000, 6000.0, 6000, 6000.

d. 34/3 = 11.333333 rounds to 11.333, 11.3, 10, 0.

e. 578.555 rounds to 578.555, 578.6, 580, 600.

f. 0.4523768 rounds to 0.452, 0.5, 0, 0.

g. −12.1 rounds to −12.100, −12.1, −10, 0.

h. −850.7654 rounds to −850.765, −850.8, −850, −900.

i. −10,995.6239 rounds to −10,995.624, −10,995.6, −11,000, −11,000.

9. Accuracy and Precision. The first scale has better accuracy since it produces a measurement that is closer to the exact weight. The second scale has better precision since it has better resolution in its measurement. While it

may seem strange, better precision may not always mean better accuracy, if other sources of error are present.

11. Counting Significant Digits.

a. The number 401 has three significant digits (because the zero is between non-zeros). It is precise to the nearest person.

b. The number 200.0 has four significant digits (because the rightmost zero is significant making the other two zeros significant as well). It is precise to the nearest tenth (0.1) of a liter.

c. The number 1.00098 has six significant digits (because the zero is between non-zeros). It is precise to the nearest 0.00001 mm.

d. The number 0.000202 has three significant digits (because it can be written 2.02×10^{-4}). It is precise to the nearest 0.000001 meters.

13. Ambiguity in Zeros.

a. If a measurement were 300,000 to the nearest hundred, it could be interpreted as $300,000 \pm 50$, and it would have four significant digits.

b. If a measurement were 300,000 to the nearest thousand, it would have three significant digits.

c. If a measurement were 300,000 to the nearest ten thousand, it would have two significant digits.

d. If a measurement were 300,000 to the nearest hundred thousand, it would have one significant digit.

15. Removing Ambiguity with Scientific Notation. The leading number in the scientific notation should have the specified number of significant digits.

a. 5×10^5

b. 5.0×10^5

c. 5.0000×10^5

d. 5.0000000×10^5.

17. Confidence and Implied Uncertainty.

a. This height measurement implies five significant digits and accuracy to the nearest ten thousandth of an inch.

This precision is not justified in a measurement of someone's height.

b. The statement implies that the votes on both sides were counted to the nearest vote. While large election results may be accurate to the nearest ten or hundred, there are too many sources of error to achieve accuracy to the nearest vote. Otherwise, we wouldn't have recounts.

c. The range of uncertainty in this estimate places the price of textbooks between $70 and $170. This is a reasonable range for textbooks prices.

d. The population figure 260 million implies a range uncertainty of 255 million to 265 million. This is a reasonable range and level of error for a U.S. census.

e. Predicting a Super Bowl victory in 5 years, implies an accuracy of ±0.5 year, which means a Super Bowl victory within exactly 5 years. This precision is not warranted.

19. Rounding Rule for Addition and Subtraction.
Recall an answer obtained by adding or subtracting two approximate numbers should be rounded to the same precision as the *least* precise of the two numbers.

a. The distance 36 cm is the least precise (to the nearest unit). If we round the sum of the given numbers (44.22 cm) to the nearest unit, we have 44 cm.

b. The weight 260 kg is the least precise (to the nearest ten). If we round the sum of the given numbers (277 kg) to the nearest ten, we have 280 kg.

c. The volume 140 liters is the least precise (to the nearest ten). If we round the difference of the given numbers (138.91 liters) to the nearest ten, we have 140 liters.

d. The time 2 hours, 37 minutes (which we will write as 2:37) is the least precise (to the nearest minute). If we round the difference of the given numbers (1:14:45) to the nearest minute, we have 1:15.

e. The distance 4.093×10^{10} km or 409.3×10^8 km is expressed to the nearest 10^7. If we round the sum of the given numbers (415.4×10^8 km) to the nearest 10^7, we have 415.4×10^8 km.

f. The weight 72 kilograms is the least precise (rounded to the nearest kilogram). If we round the difference of the given numbers (68.5) to the nearest kilogram, we have 69 kilograms.

21. Rounding Rule for Multiplication and Division.

a. The measurement 105 meters has three significant digits. The measurement 26 meters has two significant digits. The product of the given numbers is 2730 m² which should be rounded to two significant digits or 2700 m².

b. The measurement 110 km has two significant digits. The time 55 minutes has two significant digits. The quotient of the given numbers is 2.0, rounded to two significant digits.

c. The weight 9.7 kg has two significant digits. The weight 165 kg has three significant digits. The product of the given numbers is 1600.5 kg² which should be rounded to two significant digits, or 1600 kg².

d. The weight 5 gm has one significant digit. The volume 1.3 cm³ has two significant digits. The quotient of the given numbers is 3.85 gm/cm³ which should be rounded to one significant digit, or 4 gm/cm³.

5. FINANCIAL MANAGEMENT

Overview

One of the most immediate ways in which mathematics affects every person's everyday life is through finances: bank accounts, credit cards, loans, and investments. In this chapter we take an in-depth look at financial matters. As you will see there are many relevant topics in this chapter, and plenty of good mathematics.

Unit 5A The Power of Compound Interest

Overview

The phenomenon of **compounding** plays a fundamental role in all of finance, both on the investment side of the coin (earning money) and on the loan side (borrowing money). The unit opens by considering simple and compound interest problems as they arise in banking. Compound interest problems rely on four pieces of information:

- **initial deposit**, which we call P,
- **annual percentage interest rate**, which we call APR,
- **number of compoundings per year**, which we call n, and
- **number of years** the account is held, which we call Y.

You will see that there are precise formulae that tell you how much a bank account increases in value if you know these four pieces of information. These formulae (and others in the chapter) can get rather complicated and must be evaluated on a calculator. Be sure you study the highlighted boxes that show you how to use your calculator on these formulae; needless to say, practice helps immensely!

An important distinction must be made between the APR and the **annual percentage yield** (APY) of a bank account. If an account uses compounding at more than once a year, then the balance will increase by *more* than the APR in one year (due to the power of compounding). The amount by which the balance in the account actually increases in a year is the APY.

As you will see, the more often compounding takes place during the year, the more the balance increases. The limiting case occurs when compounding takes place every instant, or **continuously**. With continuous compounding, you get the maximum return on your money (for a fixed APR). The continuous compounding formula involves the mathematical constant e which is approximately 2.71828. You should become familiar with how to compute with e on your particular calculator.

The unit concludes with a very practical problem. The usual compound interest formulae tell you how much money you will have in your account after Y years with a given initial deposit *today*. But what if you know you would like to have, say $30,000, in 20 years? How much should you deposit *today* in order to reach this goal? This is an example of a present value problem value, and it's quite important for planning purposes.

Key Words and Phrases

simple interest	principal	compound interest
annual percentage rate	continuous compounding	compound interest formula
annual percentage yield		

Key Concepts and Skills

- determine the balance in an account with simple interest with a given initial deposit and interest rate.
- determine the balance in an account with compounding with a given initial deposit, *APR*, number of compoundings, and number of years.
- determine the balance in an account with continuous compounding with a given initial deposit, *APR*, and number of years.
- determine the present value for an account that will yield a given balance after a specified number of years.

Solutions Unit 5A

1. Simple vs. Compound Interest. With simple interest, the interest payment is based on the initial balance and remains constant each year. With compound interest, the interest payment is based on the current balance and increases each year.

Year	You Interest	You Balance	Your friend Interest	Your friend Balance
0	–	$500	–	$500
1	$25	$525	$25	$525
2	$25	$550	$26.25	$551.25
3	$25	$575	$27.56	$578.81
4	$25	$600	$28.94	$607.75
5	$25	$625	$30.39	$638.14

Notice that the balance grows more quickly with compound interest

3. Calculating Compound Interest. We must use the following formula.

$$A = P \times (1 + APR)^Y, \quad \text{where} \begin{cases} P = \text{starting principal} \\ APR = \text{annual percentage rate} \\ Y = \text{number of years} \\ A = \text{accumulated balance} \end{cases}$$

a. Setting $P = \$2000$, $APR = 0.03$, and $Y = 10$, we find that the accumulated balance is

$$A = P \times (1 + APR)^Y = \$2000 \times (1.03)^{10} = \$2687.83.$$

b. Setting $P = \$10,000$, $APR = 0.05$, and $Y = 20$, we find that the accumulated balance is

$$A = P \times (1 + APR)^Y = \$10,000 \times (1.05)^{20} = \$26,532.98.$$

c. Setting $P = \$30,000$, $APR = 0.07$, and $Y = 25$, we find that the accumulated balance is

$$A = P \times (1 + APR)^Y = \$30,000 \times (1.07)^{25} = \$162,822.98.$$

5. Compounding More Than Once a Year. For this problem, we use the following compound interest formula.

$$A = P\left(1 + \frac{APR}{n}\right)^{nY}, \quad \text{where} \begin{cases} P = \text{starting principal} \\ APR = \text{annual percentage rate} \\ n = \text{compoundings per year} \\ Y = \text{number of years} \\ A = \text{accumulated balance.} \end{cases}$$

a. Setting $P = \$1000$, $APR = 0.055$, $Y = 10$, and $n = 12$ (for monthly compounding), the accumulated balance is

$$A = P\left(1 + \frac{APR}{n}\right)^{nY} = \$1000\left(1 + \frac{0.055}{12}\right)^{12 \times 10} = \$1731.08.$$

b. Setting $P = \$2000$, $APR = 0.03$, $Y = 5$, and $n = 365$ (for daily compounding), the accumulated balance is

$$A = P\left(1 + \frac{APR}{n}\right)^{nY} = \$2000\left(1 + \frac{0.03}{365}\right)^{365 \times 5} = \$2323.65.$$

c. a. Setting $P = \$5000$, $APR = 0.073$, $Y = 20$, and $n = 4$ (for quarterly compounding), the accumulated balance is

$$A = P\left(1 + \frac{APR}{n}\right)^{nY} = \$5000\left(1 + \frac{0.073}{4}\right)^{4 \times 20} = \$21,248.27.$$

d. Setting $P = \$10,000$, $APR = 0.062$, $Y = 5$, and $n = 12$ (for monthly compounding), the accumulated balance is

$$A = P\left(1 + \frac{APR}{n}\right)^{nY} = \$10,000\left(1 + \frac{0.062}{12}\right)^{12 \times 5} = \$13,623.37.$$

Compound Interest and Annual Yield.

7. With $P = \$500$, $APR = 0.065$, and $n = 365$, the compound interest formula gives the following balance after $Y = 1$ year.

$$A = P\left(1 + \frac{APR}{n}\right)^{nY} = \$500\left(1 + \frac{0.065}{365}\right)^{365 \times 1} = \$533.58.$$

The annual yield is the percentage increase in the balance in one year. We find that

$$\text{Yield} = \frac{\$533.58 - \$500}{\$500} = 0.067 = 6.7\%.$$

The account has increased its value by 6.7% in one year.

9. With $P = \$800$, $APR = 0.0725$, and $n = 4$, the compound interest formula gives the following balance after $Y = 1$ year.

$$A = P\left(1 + \frac{APR}{n}\right)^{nY} = \$800\left(1 + \frac{0.0725}{4}\right)^{4 \times 1} = \$859.60.$$

The annual yield is the percentage increase in the balance in one year. We find that

$$\text{Yield} = \frac{\$859.60 - \$800}{\$800} = 0.075 = 7.5\%.$$

The account has increased its value by 7.5% in one year.

11. Annual Yield. We will use the formula for APY given in Problem 10.

a. With $APR = 0.066$ and $n = 4$, the APY is

$$APY = \left(1 + \frac{APR}{n}\right)^n - 1 = \left(1 + \frac{0.066}{4}\right)^4 - 1 = 0.0677.$$

The APY is 6.77%.

b. With $APR = 0.066$ and $n = 12$, the APY is

$$APY = \left(1 + \frac{APR}{n}\right)^n - 1 = \left(1 + \frac{0.066}{12}\right)^{12} - 1 = 0.0680.$$

The APY is 6.80%.

c. With $APR = 0.066$ and $n = 365$, the APY is

$$APY = \left(1 + \frac{APR}{n}\right)^n - 1 = \left(1 + \frac{0.066}{365}\right)^{365} - 1 = 0.0682.$$

The APY is 6.82%.

Notice that in all cases the APY is greater than the APR.

13. Rates of Compounding. For Account 1, we set $P = \$1000$, $APR = 0.055$, and $n = 1$ in the compound interest formula. For Account 2, we set $P = \$1000$, $APR = 0.055$, and $n = 365$ in the compound interest formula. We let Y vary from 1 to 10. All figures are rounded to the nearest dollar.

	Account 1		Account 2	
Year	Interest	Balance	Interest	Balance
0	–	$1000	–	$1000
1	$55	$1055	$57	$1057
2	$58	$1113	$59	$1116
3	$61	$1174	$63	$1179
4	$65	$1239	$67	$1246
5	$68	$1307	$70	$1317
6	$72	$1379	$74	$1391
7	$76	$1455	$79	$1470
8	$80	$1535	$83	$1553
9	$84	$1619	$87	$1640
10	$89	$1708	$93	$1733

Account 1 has increased in value by $708 or 70.8%. Account 2 has increase in value by $733 or 73.3%. The difference in the balance of the two accounts is due to the number of compoundings per year.

15. Continuous Compounding. Recall the continuous compounding formula:

$$A = P \times e^{(APR \times Y)}, \quad \text{where} \begin{cases} P = \text{starting principal} \\ APR = \text{annual percentage rate} \\ Y = \text{number of years} \\ A = \text{accumulated balance after } Y \text{ years} \end{cases}$$

a. With $P = \$1000$, and $APR = 0.04$, we find that the accumulated balance after $Y = 5$ years is

$$A = P \times e^{(APR \times Y)} = \$1000 \times e^{0.04 \times 5} = \$1221.40.$$

After $Y = 20$ years, the accumulated balance is

$$A = P \times e^{(APR \times Y)} = \$1000 \times e^{0.04 \times 20} = \$2225.54.$$

b. With $P = \$2000$, and $APR = 0.05$, we find that the accumulated balance after $Y = 5$ years is

$$A = P \times e^{(APR \times Y)} = \$2000 \times e^{0.05 \times 5} = \$2568.05.$$

After $Y = 20$ years, the accumulated balance is

$$A = P \times e^{(APR \times Y)} = \$2000 \times e^{0.05 \times 20} = \$5436.56.$$

c. With $P = \$10,000$, and $APR = 0.06$, we find that the accumulated balance after $Y = 5$ years is

$$A = P \times e^{(APR \times Y)} = \$10,000 \times e^{0.06 \times 5} = \$13,498.59.$$

After $Y = 20$ years, the accumulated balance is

$$A = P \times e^{(APR \times Y)} = \$10,000 \times e^{0.06 \times 20} = \$33,201.17.$$

17. APY for Continuous Compounding. Because the APY is independent of the initial balance, we can take $P = \$100$ (or any number) to determine how much the balance increases in one year.

a. With $APR = 0.04$, an initial deposit of $P = \$100$ will grow in one year to

$$A = P \times e^{(APR \times Y)} = \$100 \times e^{0.04 \times 1} = \$104.08.$$

The account has increased in value by $4.08, so the APY is $4.08/$100 = 4.08%.

b. With $APR = 0.05$, an initial deposit of $P = \$100$ will grow in one year to

$$A = P \times e^{(APR \times Y)} = \$100 \times e^{0.05 \times 1} = \$105.13.$$

The account has increased in value by $5.13, so the APY is $5.13/$100 = 5.13%.

c. With $APR = 0.06$, an initial deposit of $P = \$100$ will grow in one year to

$$A = P \times e^{(APR \times Y)} = \$100 \times e^{0.06 \times 1} = \$106.18.$$

The account has increased in value by $6.18, so the APY is $6.18/$100 = 6.18%.

P 19. Continuous Compounding.

a. We must fill in the table entries using the formula

$$APY = \left(1 + \frac{APR}{n}\right)^n - 1,$$

with $APR = 0.12$, for various values of n. Here is the result.

n	1	4	12	365	500	1000
APY	12.00	12.55	12.68	12.75	12.75	12.75

We see that compounding more than once a day (365 times a year) has very little effect.

b. With continuous compounding, the APY would be

$$APY = e^{0.12} - 1 = 0.1275 = 12.75\%.$$

The APY with continuous compounding is identical (to the nearest hundredth) to the APY with daily compounding.

e. With $P = \$500$, and $APR = 0.12$, we find that the accumulated balance after $Y = 1$ year is

$$A = P \times e^{(APR \times Y)} = \$500 \times e^{0.12 \times 1} = \$563.75.$$

After $Y = 5$ years, the accumulated balance is

$$A = P \times e^{(APR \times Y)} = \$500 \times e^{0.12 \times 5} = \$911.06.$$

21. Comparing Investment Plans. With $APR = 4.5\%$ and an initial deposit of $P = \$1600$, after $Y = 5$ years, Brian will have an accumulated balance of

$$A = P \times e^{(APR \times Y)} = \$1600 \times e^{0.045 \times 5} = \$2003.72.$$

After $Y = 20$ years, his accumulated balance will be

$$A = P \times e^{(APR \times Y)} = \$1600 \times e^{0.045 \times 20} = \$3935.36.$$

With $APR = 5.5\%$ and an initial deposit of $P = \$1400$, after $Y = 5$ years, Celeste will have an accumulated balance of

$$A = P \times e^{(APR \times Y)} = \$1400 \times e^{0.055 \times 5} = \$1843.14.$$

After $Y = 20$ years, her accumulated balance will be

$$A = P \times e^{(APR \times Y)} = \$1400 \times e^{0.055 \times 20} = \$4205.83.$$

We see that after five years, Brian, with the largest initial balance, has the best investment. However, after twenty years, Celeste, with the highest interest rate, has the best investment.

23. Planning Ahead With Compounding.

a. In this problem we must find that initial deposit P that results in an accumulated balance of $10,000. We have $n = 1$ (annual compounding), $Y = 10$, and $APR = 9\%$. The initial deposit P must satisfy

$$\$10,000 = P \times (1+0.09)^{10} = P \times 2.3674.$$

We can solve this directly for P by dividing both sides of the equation by 2.3673. The required initial deposit is

$$P = \frac{\$10,000}{2.3674} = \$4224.$$

An initial deposit of $4224 is needed to accumulate $10,000 in ten years with annual compounding.

b. We now have $n = 4$ (quarterly compounding), $Y = 10$, and $APR = 9\%$. The initial deposit P must satisfy

$$\$10,000 = P\left(1+\frac{0.09}{4}\right)^{4 \times 10} = P \times (1.0225)^{40} = P \times 2.4352.$$

We can solve this directly for P by dividing both sides of the equation by 2.4352. The required initial deposit is

$$P = \frac{\$10,000}{2.4352} = \$4106.$$

An initial deposit of $4106 is needed to accumulate $10,000 in ten years with quarterly compounding.

c. We have $n = 12$ (monthly compounding), $Y = 10$, and $APR = 9\%$. The initial deposit P must satisfy

$$\$10,000 = P\left(1+\frac{0.09}{12}\right)^{12 \times 10} = P \times (1.0075)^{120} = P \times 2.4514.$$

We can solve this directly for P by dividing both sides of the equation by 2.4514. The required initial deposit is

$$P = \frac{\$10,000}{2.4514} = \$4079.$$

An initial deposit of $4079 is needed to accumulate $10,000 in ten years with monthly compounding.

d. For continuous compounding, we take $Y = 10$ and $APR = 9\%$. The initial deposit P must satisfy

$$\$10,000 = Pe^{0.09 \times 10} = P \times 2.4596.$$

We can solve this directly for P by dividing both sides of the equation by 2.4596. The required initial deposit is

$$P = \frac{\$10,000}{2.4596} = \$4066.$$

An initial deposit of $4066 is needed to accumulate $10,000 in ten years with continuous compounding.

In summary, the more frequently interest in compounded, the less money is needed for an initial deposit to reach a given final balance.

25. Retirement Planning.

The target accumulated balance is $A = \$75,000$. For Plan A we must use the

compound interest formula with $APR = 5\%$, $Y = 35$ years, and $n = 1$ (annual compounding). We want to find the initial deposit P that satisfies

$$\$75,000 = P \times (1+0.05)^{35} = P \times 5.5160.$$

We can solve this directly for P by dividing both sides of the equation by 5.5160. The required initial deposit is

$$P = \frac{\$75,000}{5.5160} = \$13,597.$$

An initial deposit of $13,597 is needed to accumulate $75,000 with Plan A.

For Plan B we must use the continuous compound interest formula with $APR = 4.5\%$ and $Y = 35$ years. We want to find the initial deposit P that satisfies

$$\$75,000 = P \times e^{0.045 \times 35} = P \times 4.8307.$$

We can solve this directly for P by dividing both sides of the equation by 4.8307. The required initial deposit is

$$P = \frac{\$35,000}{4.8307} = \$15,526.$$

An initial deposit of $15,526 is needed to accumulate $75,000 with Plan B. Plan A requires the smaller initial deposit to achieve the goal of $75,000 in 35 years.

P 29. Working With the Compound Interest Formula.

a. The tripling time for an account will be the same regardless of the initial deposit. So let's take $P = \$100$ (any other number will also work) and find the value of Y that gives an accumulated balance of $A = \$300$, with $n = 1$ and $APR = 8\%$. The problem can be solved exactly using logarithms, but we will proceed by trial and error. Using the interest formula for more than one compounding per year, we must find the value of Y that makes the following statement true:

$$\$300 = \$100(1+0.08)^Y.$$

Substituting $Y = 10$, we find that

$$\$100(1+0.08)^{10} = \$216,$$

which is less than $300. So a larger value of Y is needed. Substituting $Y = 15$, we find that

$$\$100(1+0.08)^{15} = \$317,$$

which is more than $300. Continuing in this way, you can check that a value of $Y = 14.3$ satisfies the equation. Therefore the tripling time for the account is about 14.3 years.

c. In this case, we are looking for the time needed for the balance to increase 100-fold. Setting $APR = 7\%$, $n = 1$, $P = \$1000$, and $A = \$100,000$, we must find the value of Y that satisfies

$$\$100,000 = \$1000(1+0.07)^Y.$$

Some trial and error (or logarithms) shows that the required time is about 68.1 years.

Unit 5B Savings Plans (Annuities)

Overview

An **annuity** is a special kind of savings plan in which deposits are made regularly, perhaps every month or every year. People who are creating a college fund or building a retirement plan usually use an annuity. An annuity account increases in value due to compounding (as in regular bank accounts as studied in the previous unit) and due to the regular deposits. Not surprisingly, the mathematics of annuity plans follows naturally from the compound interest formulae of the previous unit.

There is just one basic formula that needs to be mastered in order to work with annuities. The formula requires all of the input needed for the compound interest formula of the previous unit (initial deposit, APR, number of compoundings, and number of years), *plus* the amount of the regular deposits. To keep matters simple (but still realistic) we assume that the regular deposits are made as often as interest is compounded. There is no doubt that the savings plan formula is complicated. Be sure to study the highlighted box to learn how your calculator can be used to evaluate the formula.

Just as with a bank account, it is practical to ask present value questions with savings plans. If you know how much money you would like to have at a future time (perhaps the time of retirement), then how much should you deposit, say monthly, between now and then in order to reach that goal? You will see some practical examples of solving present value problems.

Key Words and Phrases

savings plan   annuity   savings plan formula

Key Concepts and Skills

- determine the value of a savings plan given the initial deposit, APR, number of compoundings, number of years, and the amount of the regular deposits
- determine the present value for a savings plan that will yield a given balance after a specified number of years.

Solutions Unit 5B

Investment Plans. In the next four problems, we will use the savings plan formula:

$$A = PMT \times \left[\frac{\left(1 + \frac{APR}{n}\right)^{nY} - 1}{\frac{APR}{n}} \right],$$

where A = accumulated balance, PMT = regular payment amount, APR = annual percentage rate (as a decimal), n = number of payments per year, and Y = number of years.

1. To model this IRA, set $PMT = \$50$, $APR = 0.08$, $n = 12$ payments per year, and $Y = 40$ years. The accumulated balance at age 65 will be

$$A = \$50 \times \left[\frac{\left(1 + \frac{0.08}{12}\right)^{12 \times 40} - 1}{\frac{0.08}{12}} \right] = \$174,550.39.$$

The total amount deposited into the account over the 40-year period is

$$\frac{\$50}{\text{month}} \times \frac{12 \text{ months}}{\text{year}} \times 40 \text{ years} = \$24,000.$$

Thus the IRA has earned about $150,000 in interest.

3. In this case, we set $PMT = \$200$, $APR = 0.07$, $n = 12$ payments per year, and $Y = 18$ years. The accumulated balance after 18 years will be

$$A = \$200 \times \left[\frac{\left(1 + \frac{0.07}{12}\right)^{12 \times 18} - 1}{\frac{0.07}{12}} \right] = \$86,144.21.$$

The total amount deposited into the account over the 18-year period is

$$\frac{\$200}{\text{month}} \times \frac{12 \text{ months}}{\text{year}} \times 18 \text{ years} = \$43,200.$$

About half of the IRA was earned through interest.

Who Comes Out Ahead?

5. We use the savings plan formula (given in Problem 1 above). For Yolanda's investment plan, we set $PMT = \$100$, $APR = 0.05$, and $n = 12$ payments per year. The accumulated balance after $Y = 10$ years will be

$$A = \$100 \times \left[\frac{\left(1 + \frac{0.05}{12}\right)^{12 \times 10} - 1}{\frac{0.05}{12}} \right] = \$15,528.23.$$

Yolanda's total payments are

$$\frac{\$100}{\text{month}} \times \frac{12 \text{ months}}{\text{year}} \times 10 \text{ years} = \$12,000.$$

For Zach's investment plan, we set $PMT = \$1200$, $APR = 0.05$, and $n = 1$ payment per year. The accumulated balance after $Y = 10$ years will be

$$A = \$1200 \times \left[\frac{\left(1 + \frac{0.05}{1}\right)^{1 \times 10} - 1}{\frac{0.05}{1}} \right] = \$15,093.47.$$

Zach's total payments are $1200/yr × 10 yr = $12,000. Although Yolanda and Zach deposit the same amount into their plans, Yolanda has the larger balance after 10 years

because of the more frequent payments and compounding.

7. We use the savings plan formula (given in Problem 1 above). For Juan's investment plan, we set $PMT = \$200$, $APR = 0.06$, and $n = 12$ payments per year. The accumulated balance after $Y = 10$ years will be

$$A = \$200 \times \left[\frac{\left(1 + \frac{0.06}{12}\right)^{12 \times 10} - 1}{\frac{0.06}{12}} \right] = \$32,775.87.$$

Juan's total payments are

$$\frac{\$200}{\text{month}} \times \frac{12 \text{ months}}{\text{year}} \times 10 \text{ years} = \$24,000.$$

For Maria's investment plan, we set $PMT = \$2500$, $APR = 0.065$, and $n = 1$ payment per year. The accumulated balance after $Y = 10$ years will be

$$A = \$2500 \times \left[\frac{\left(1 + \frac{0.065}{1}\right)^{1 \times 10} - 1}{\frac{0.065}{1}} \right] = \$33,736.06.$$

Maria's total payments are $2500/yr × 10 yr = $25,000. Maria deposits more into her savings plan and has a higher APR than Juan over the long run. Despite the fact that her account compounds interest only once a year, she still comes out ahead.

Investment Planning. In Problems 10–14, we will use the Savings Plan Formula Solved for Payments:

$$PMT = A \times \left[\frac{\frac{APR}{n}}{\left(1 + \frac{APR}{n}\right)^{nY} - 1} \right]$$

9. We set $A = \$150,000$, $Y = 18$ years, $APR = 0.075$, and $n = 12$ (for monthly deposits) in the Savings Plan Formula Solved for Payments. We find that the required monthly payments are

$$PMT = \$150,000 \times \left[\frac{\frac{0.075}{12}}{\left(1 + \frac{0.075}{12}\right)^{12 \times 18} - 1} \right] = \$329.96.$$

Notice that you pay only

$$\frac{\$329.96}{\text{month}} \times \frac{12 \text{ months}}{\text{year}} \times 18 \text{ years} = \$71,271.36,$$

but earn more than twice this amount.

11. We set $A = \$10,000$, $Y = 3$ years, $APR = 0.055$, and $n = 12$ (for monthly deposits) in the Savings Plan Formula Solved for Payments. We find that the required monthly payments are

$$PMT = \$10,000 \times \left[\frac{\frac{0.055}{12}}{\left(1 + \frac{0.055}{12}\right)^{12 \times 3} - 1} \right] = \$256.13.$$

Notice that you pay only

$$\frac{\$256.13}{\text{month}} \times \frac{12 \text{ months}}{\text{year}} \times 3 \text{ years} = \$9220.68.$$

13. In this problem we are asked to find the number of years, Y, required for the accumulated balance to reach A = \$50,000, with monthly deposits ($n = 12$) of PMT =\$100 and $APR = 0.07$. We begin with the savings plan formula and substitute values:

$$\$50,000 = \$100 \times \left[\frac{\left(1 + \frac{0.07}{12}\right)^{12 \times Y} - 1}{\frac{0.07}{12}} \right]$$

At this point, we could solve this equation for the unknown Y (using some algebra and logarithms). Instead, we will proceed by trial and error, substituting various values of Y, in hopes of satisfying the equation. For example, if we set $Y = 15$ years, then the accumulated balance is

$$A = \$100 \times \left[\frac{\left(1 + \frac{0.07}{12}\right)^{12 \times 15} - 1}{\frac{0.07}{12}} \right] = \$31,696.23,$$

which is less than the desired balance of \$50,000. If we set $Y = 20$ years, then the accumulated balance is

$$A = \$100 \times \left[\frac{\left(1 + \frac{0.07}{12}\right)^{12 \times 20} - 1}{\frac{0.07}{12}} \right] = \$52,092.67,$$

which is slightly higher than the desired balance of \$50,000. Proceeding in this manner, you can get better and better approximations to the exact value of Y. Eventually, you would discover that $Y = 19.56$ years, gives a balance very near \$50,000. Similarly, you can determine that it takes about $Y = 58$ years to accumulate a balance of \$1,000,000.

Comparing Investment Plans.

15. Using the savings plan formula with $PMT = \$75$, $n = 12$ (monthly payments), $APR = 0.07$, and $Y = 15$, we can compute an accumulated balance of

$$A = \$75 \times \left[\frac{\left(1 + \frac{0.07}{12}\right)^{12 \times 15} - 1}{\frac{0.07}{12}} \right] = \$23,772.17,$$

which is short of the goal of \$50,000.

17. Comfortable Retirement. As in Example 3 of the text, there are two parts to this problem. The first part is to determine how much money is needed at the time of retirement so that you can live *entirely* on the interest on the annuity; that is, the interest on the annuity must be \$50,000 per year. If we assume that interest is compounded annually at $APR = 0.08$ during the pay-out period, the accumulated balance in the annuity must be such that

$$\$50,000 = 8\% \times (\text{needed balance})$$

Dividing both sides by $8\% = 0.08$ we find:

$$\text{needed balance} = \frac{\$50,000}{0.08} = \$625,000.$$

The second part is to determine how much you must pay into the annuity while you are working to reach the goal of \$625,000 at age 60. We set $A = \$625,000$, $Y = 30$ years, $APR = 0.08$, and $n = 12$ (for monthly deposits) in the Savings Plan Formula Solved for Payments. We find that the required monthly payments are

$$PMT = \$625,000 \times \left[\frac{\frac{0.08}{12}}{\left(1 + \frac{0.08}{12}\right)^{12 \times 30} - 1} \right] = \$419.36.$$

If you pay \$419.36 monthly between the ages of 30 and 60, you will be able to retire and earn \$50,000 per year – forever. This calculation makes the assumption that interest rates remain at 8% during both the pay-in and pay-out phases of the plan.

19. Regular Deposits vs. Lump Sum.

a. In Unit 5A we used the compound interest formula to determine how much must be deposited once (in a lump sum) into an account to earn a certain accumulated balance. With an initial deposit of A = \$60,000, $APR = 0.045$, $n = 12$ (monthly compounding), and $Y = 20$ years the lump sum deposit is

$$P = \frac{A}{\left(1 + \frac{APR}{n}\right)^{nY}} = \frac{\$60,000}{\left(1 + \frac{0.045}{12}\right)^{12 \times 20}} = \$24,435.28.$$

A single lump sum deposit of \$24,435.28 must be made to earn \$60,000 in 20 years.

21 Variable Rates.

a. After $Y = 2$ years with $APR = 0.06$ and monthly payments of $PMT = \$100$, your accumulated balance will be

$$A = \$100 \times \left[\frac{\left(1 + \frac{0.06}{12}\right)^{12 \times 2} - 1}{\frac{0.06}{12}} \right] = \$2543.20 .$$

b. We can now regard this balance of \$2543.20 as an initial (lump sum) deposit into a new account with monthly payments of \$125. As in the previous problem, the accumulated balance after $Y = 5$ years will consist of the contribution from the initial deposit and the contribution from the monthly payments.

From the lump sum deposit of $P = \$2543.20$ with $APR = 0.06$, $n = 12$, and $Y = 5$, the accumulated balance is

$$A = P\left(1 + \frac{APR}{n}\right)^{nY} = \$2543.20\left(1 + \frac{0.06}{12}\right)^{12 \times 5} = \$3430.40.$$

From the monthly payments with $PMT = \$125$, $APR = 0.06$, $n = 12$, and $Y = 5$, the accumulated balance is

$$A = \$125 \times \left[\frac{\left(1 + \frac{0.06}{12}\right)^{12 \times 5} - 1}{\frac{0.06}{12}} \right] = \$8721.25 .$$

After an additional five years, the account will hold $\$3430.40 + \$8721.25 = \$12,151.65$

Unit 5C Loan Payments, Credit Cards, and Mortgages

Overview

While most people have bank accounts or investments that earn money, the sad reality is that most people also have debts due to loans of one kind or another. In this unit we will explore the most common kinds of loans: short-term loans (such as automobile loans), loans on credit cards, and **mortgages** (or house loans).

One basic formula governs all loan problems; it is called the **loan payment formula**. Given

- the amount of the loan (called the **principal**),
- the interest rate on the loan (still called the *APR*),
- the number of payment periods per year (usually taken to be 12 for monthly payments), and
- the term of the loan (the number of years over which it will be paid back),

the loan payment formula tells you the amount of the regular payments on the loan. Once again, the loan payment formula is rather complicated, so be sure to study the highlighted box on using your calculator.

The entire unit is devoted to using the loan payment formula for various practical problems. Often you must choose between a loan with a high interest rate and a short term and a loan with a lower interest rate but a longer term. Which is the best choice? You will see several examples of such decisions.

The biggest loan that many people will ever use is a house loan or **mortgage**. Since mortgages involve such large amounts of money over long periods of time, the decisions you make in choosing a mortgage are critical. We explore some of the options and strategies involved with mortgages. Specifically, we explain the differences between fixed-rate mortgages as opposed to adjustable rate mortgages. We also discuss issues of prepayments, refinancing, points, and closing costs. All in all, you will find this to be quite a practical unit.

Key Words and Phrases

loan principal	installment loan	loan payment formula
mortgage	down payment	fixed rate mortgage
adjustable rate mortgage	closing costs	points
prepayment penalties	refinance	

Key Concepts and Skills
- know the terminology associated with loans.
- determine the loan payment given the principal, the APR, the number of payment periods, and the term of the loan.
- analyze two loan options to determine which is best for a given situation.

Solutions Unit 5C

1. Loan Terminology.

a. The principal of the loan is $40,000.

b. The APR is 7%.

c. The monthly payments are $310.

d. The term of the loan is 20 years.

e. The number of monthly payments is (12 payments/yr) × 20 yr = 240 payments.

f. The total amount paid out over the term of the loan is ($310/payment) × 240 payments = $74,400.

g. Because the principal of the loan is only $40,000, you will pay $74,400 − $40,000 = $34,400 as interest. This means that $34,400/$74,400 = 46% of your payments will be interest.

3. Loan Payments. We need to use the loan payment formula for these problems:

$$PMT = \frac{P \times \left(\frac{APR}{n}\right)}{1 - \left(1 + \frac{APR}{n}\right)^{-nY}}, \quad \text{where} \begin{cases} P = \text{starting loan principal} \\ PMT = \text{regular payment amount} \\ APR = \text{annual percentage rate} \\ n = \text{payment periods per year} \\ Y = \text{loan term in years.} \end{cases}$$

a. We set $P = \$25,000$, $APR = 0.10$, $n = 12$ (for monthly payments) and $Y = 20$ years in the loan payment formula. The monthly payments are

$$PMT = \frac{P \times \left(\frac{APR}{n}\right)}{1 - \left(1 + \frac{APR}{n}\right)^{-nY}} = \frac{\$25,000 \times \left(\frac{0.10}{12}\right)}{1 - \left(1 + \frac{0.10}{12}\right)^{-12 \times 20}} = \$241.26.$$

b. We set $P = \$150,000$, $APR = 0.095$, $n = 12$ (for monthly payments) and $Y = 30$ years in the loan payment formula. The monthly payments are

$$PMT = \frac{P \times \left(\frac{APR}{n}\right)}{1 - \left(1 + \frac{APR}{n}\right)^{-nY}} = \frac{\$150,000 \times \left(\frac{0.095}{12}\right)}{1 - \left(1 + \frac{0.095}{12}\right)^{-12 \times 30}} = \$1261.28.$$

c. We set $P = \$150,000$, $APR = 0.0875$, $n = 12$ (for monthly payments) and $Y = 15$ years in the loan payment formula. The monthly payments are

$$PMT = \frac{P \times \left(\frac{APR}{n}\right)}{1 - \left(1 + \frac{APR}{n}\right)^{-nY}} = \frac{\$150,000 \times \left(\frac{0.0875}{12}\right)}{1 - \left(1 + \frac{0.0875}{12}\right)^{-12 \times 15}} = \$1499.17.$$

5. Monthly Payments. We must use the loan payment formula given above in the solution to Problem 3. Setting $P = \$5000$, $APR = 0.12$, $n = 12$ (for monthly payments) and $Y = 3$ years in the loan payment formula, we find that the monthly payments are

$$PMT = \frac{P \times \left(\frac{APR}{n}\right)}{1 - \left(1 + \frac{APR}{n}\right)^{-nY}} = \frac{\$5000 \times \left(\frac{0.12}{12}\right)}{1 - \left(1 + \frac{0.12}{12}\right)^{-12 \times 3}} = \$166.07.$$

The total payments over the term of the loan are $166.07/payment × 36 payments = $5978.52. Of the total payments, $5000 goes to the principal and $5978.52 − $5000 = $978.52 is interest. This means that $978.52/$5978.52 = 16% of the payments are interest.

7. Monthly Payments. We must use the loan payment formula given above in the solution to Problem 3. Setting $P = \$50,000$, $APR = 0.08$, $n = 12$ (for monthly payments) and $Y = 15$ years in the loan payment formula, we find that the monthly payments are

$$PMT = \frac{P \times \left(\frac{APR}{n}\right)}{1-\left(1+\frac{APR}{n}\right)^{-nY}} = \frac{\$50,000 \times \left(\frac{0.08}{12}\right)}{1-\left(1+\frac{0.08}{12}\right)^{-12\times15}} = \$477.83.$$

The total payments over the term of the loan are $477.83/payment × 180 payments = $86,009.40. Of the total payments, $50,000 goes to the principal and $86,009.40 − $50,000 = $36,009.40 is interest. This means that $36,009.40/$86,009.40 = 42% of the payments are interest.

9. Accelerated Student Loan Payment.

a. Using the loan payment formula with $P = \$25,000$, $APR = 0.09$, $n = 12$ (for monthly payments) and $Y = 20$ years, the required monthly payments are

$$PMT = \frac{P \times \left(\frac{APR}{n}\right)}{1-\left(1+\frac{APR}{n}\right)^{-nY}} = \frac{\$25,000 \times \left(\frac{0.09}{12}\right)}{1-\left(1+\frac{0.09}{12}\right)^{-12\times20}} = \$224.93.$$

b. If you decided to pay the loan off in $Y = 10$ years, the monthly payments would increase to

$$PMT = \frac{P \times \left(\frac{APR}{n}\right)}{1-\left(1+\frac{APR}{n}\right)^{-nY}} = \frac{\$25,000 \times \left(\frac{0.09}{12}\right)}{1-\left(1+\frac{0.09}{12}\right)^{-12\times10}} = \$316.69.$$

c. With a 20-year term, the total payments would be $224.93/payment × 240 payments = $53,985.60. With a 10-year term, the total payments would be $316.69/payment × 120 payments = $38,002.80. Not surprisingly, you pay much less with a 10-year term, but you also have to pay about $90 more each month.

11. Student Loan Consolidation.

a. First we need to compute the monthly payments for each of the three loans individually. Using the loan payment formula with $P = \$10,000$, $APR = 0.08$, $n = 12$ (for monthly payments) and $Y = 15$ years, the required monthly payment for Loan 1 is

$$PMT = \frac{P \times \left(\frac{APR}{n}\right)}{1-\left(1+\frac{APR}{n}\right)^{-nY}} = \frac{\$10,000 \times \left(\frac{0.08}{12}\right)}{1-\left(1+\frac{0.08}{12}\right)^{-12\times15}} = \$95.57.$$

Using the loan payment formula with $P = \$15,000$, $APR = 0.085$, $n = 12$ (for monthly payments) and $Y = 20$ years, the required monthly payment for Loan 2 is

$$PMT = \frac{P \times \left(\frac{APR}{n}\right)}{1-\left(1+\frac{APR}{n}\right)^{-nY}} = \frac{\$15,000 \times \left(\frac{0.085}{12}\right)}{1-\left(1+\frac{0.085}{12}\right)^{-12\times20}} = \$130.17.$$

Using the loan payment formula with $P = \$12,500$, $APR = 0.09$, $n = 12$ (for monthly payments) and $Y = 10$ years, the required monthly payment for Loan 3 is

$$PMT = \frac{P \times \left(\frac{APR}{n}\right)}{1-\left(1+\frac{APR}{n}\right)^{-nY}} = \frac{\$12,500 \times \left(\frac{0.09}{12}\right)}{1-\left(1+\frac{0.09}{12}\right)^{-12\times10}} = \$158.34.$$

b. Thus the total of all payments for the three loans is

($95.57 × 180) + ($130.17 × 240) + ($158.34 × 120)

which comes to $67,444.20.

c. The consolidated loan would have a principal equal to the sum of the principals of the three individual loans, or $10,000 + $15,000 + $12,500 = $37,500. Using the loan payment formula with $P = \$37,500$, $APR = 0.085$, $n = 12$ (for monthly payments) and $Y = 20$ years, the monthly payment for the consolidated loan is

$$PMT = \frac{P \times \left(\frac{APR}{n}\right)}{1-\left(1+\frac{APR}{n}\right)^{-nY}} = \frac{\$37,500 \times \left(\frac{0.085}{12}\right)}{1-\left(1+\frac{0.085}{12}\right)^{-12\times20}} = \$325.43.$$

Your monthly payment will be about $60 less with the consolidated loan for the first ten years. However you can verify that the total payout for the three loans would be $67,444, while the total payout for the consolidated loan would be $78,103. So if you can pay the extra $60 per month, the three separate loans would be preferable. Also, the monthly payment for the three loans decreases after 10 years (when Loan 3 is paid off) and again after 15 years (when Loan 1 is paid off).

13. Credit Card Payments.

a. Each month the new balance is found by subtracting $200 from the old balance (monthly payment), adding $75 to the old balance (new expenses), and adding the interest to the old balance. The interest is always 1.5% of the old balance.

Month 1

Balance = $1200 − $200 + $75 + (1.5% × $1200) = $1093.00.

Month 2

Balance = $1093 − $200 + $75 + (1.5% × $1093) = $984.40.

Month 3

Balance = $984.40 − $200 + $75 + (1.5% × $984.40) = $874.17.

Month 4

Balance = $874.17 − $200 + $75 + (1.5% × $874.17) = $762.28.

Month 5

Balance = $762.28 − $200 + $75 + (1.5% × $762.28) = $648.71.

Month 6

Balance = $648.71 − $200 + $75 + (1.5% × $648.71) = $533.44.

b. The loan will be paid off in the tenth month. The balances in months 7 through 10 are: $416.45, $297.69, $177.16, $54.81.

15. Credit Card Woes. With an APR of 18%, we can take the monthly interest rate to be 18%/12 = 1.5%.

Month	Payment	Expenses	Interest	Balance
0	-	-	-	$300
1	$300	$175	$4.50	$179.50
2	$150	$150	$2.69	$182.19
3	$400	$350	$2.73	$134.93
4	$500	$450	$2.02	$86.95
5	0	$100	$1.30	$188.25
6	$100	$100	$2.82	$191.08
7	$200	$150	$2.87	$143.94
8	$100	$80	$2.16	$126.10

The balance is never eliminated even though payments equal or exceed expenses in all but one month.

17. Choosing an Auto Loan. Let's look at the monthly payments for each loan.

Using the loan payment formula with $P = \$10,000$, $APR = 0.07$, $n = 12$ (for monthly payments) and $Y = 3$ years, the monthly payment for Loan 1 is

$$PMT = \frac{P \times \left(\frac{APR}{n}\right)}{1 - \left(1 + \frac{APR}{n}\right)^{-nY}} = \frac{\$10,000 \times \left(\frac{0.07}{12}\right)}{1 - \left(1 + \frac{0.07}{12}\right)^{-12 \times 3}} = \$308.77.$$

Note that you end up paying a total of $\$308.77$/payment $\times$ 36 payments $= \$11,115.72$.

Using the loan payment formula with $P = \$10,000$, $APR = 0.075$, $n = 12$ (for monthly payments) and $Y = 4$ years, the monthly payment for Loan 2 is

$$PMT = \frac{P \times \left(\frac{APR}{n}\right)}{1 - \left(1 + \frac{APR}{n}\right)^{-nY}} = \frac{\$10,000 \times \left(\frac{0.075}{12}\right)}{1 - \left(1 + \frac{0.075}{12}\right)^{-12 \times 4}} = \$241.79.$$

Note that you end up paying a total of $\$241.79$/payment $\times$ 48 payments $= \$11,605.92$.

Using the loan payment formula with $P = \$10,000$, $APR = 0.08$, $n = 12$ (for monthly payments) and $Y = 5$ years, the monthly payment for Loan 3 is

$$PMT = \frac{P \times \left(\frac{APR}{n}\right)}{1 - \left(1 + \frac{APR}{n}\right)^{-nY}} = \frac{\$10,000 \times \left(\frac{0.08}{12}\right)}{1 - \left(1 + \frac{0.08}{12}\right)^{-12 \times 5}} = \$202.76.$$

Note that you end up paying a total of $\$202.76$/payment $\times$ 60 payments $= \$12,165.60$.

Because you can afford payments of $220 per month, Loan 3 is the only option.

19. Mortgage Options. We'll compute the monthly payment for each loan.

Using the loan payment formula with $P = \$120,000$, $APR = 0.0675$, $n = 12$ (for monthly payments) and $Y = 15$ years, the monthly payment for Loan 1 is

$$PMT = \frac{P \times \left(\frac{APR}{n}\right)}{1 - \left(1 + \frac{APR}{n}\right)^{-nY}} = \frac{\$120,000 \times \left(\frac{0.0675}{12}\right)}{1 - \left(1 + \frac{0.0675}{12}\right)^{-12 \times 15}} = \$1061.89.$$

Note that you end up paying a total of $\$1061.89$/payment $\times$ 180 payments $= \$191,140.20$.

Using the loan payment formula with $P = \$120,000$, $APR = 0.07$, $n = 12$ (for monthly payments) and $Y = 20$ years, the monthly payment for Loan 2 is

$$PMT = \frac{P \times \left(\frac{APR}{n}\right)}{1 - \left(1 + \frac{APR}{n}\right)^{-nY}} = \frac{\$120,000 \times \left(\frac{0.07}{12}\right)}{1 - \left(1 + \frac{0.07}{12}\right)^{-12 \times 20}} = \$930.36.$$

Note that you end up paying a total of $\$930.36$/payment $\times$ 240 payments $= \$223,286.40$.

Using the loan payment formula with $P = \$120,000$, $APR = 0.0715$, $n = 12$ (for monthly payments) and $Y = 30$ years, the monthly payment for Loan 3 is

$$PMT = \frac{P \times \left(\frac{APR}{n}\right)}{1 - \left(1 + \frac{APR}{n}\right)^{-nY}} = \frac{\$120,000 \times \left(\frac{0.0715}{12}\right)}{1 - \left(1 + \frac{0.0715}{12}\right)^{-12 \times 30}} = \$810.49.$$

Note that you end up paying a total of $\$810.49$/payment $\times$ 360 payments $= \$291,776.40$.

This problem illustrates the dilemma with all mortgages. The 15-year mortgage requires a higher monthly payment, but has a lower total payout over the loan period. The 30-year mortgage carries a lower monthly payment, but, in the end, costs about 50% more than the 15-year mortgage. If the higher payment is affordable, Loan 1 is probably a better option. Not included in this analysis are the tax consequences of mortgage interest payments – a subject to be addressed in Unit 5D.

Closing Costs and Points.

21. We'll compute the monthly payment for each loan.

Using the loan payment formula with $P = \$80,000$, $APR = 0.08$, $n = 12$ (for monthly payments) and $Y = 30$ years, the monthly payment for Loan 1 is

$$PMT = \frac{P \times \left(\frac{APR}{n}\right)}{1 - \left(1 + \frac{APR}{n}\right)^{-nY}} = \frac{\$80,000 \times \left(\frac{0.08}{12}\right)}{1 - \left(1 + \frac{0.08}{12}\right)^{-12 \times 30}} = \$587.01.$$

You end up paying a total of $\$587.01$/payment $\times$ 360 payments $= \$211,323.60$.

Using the loan payment formula with $P = \$80,000$, $APR = 0.075$, $n = 12$ (for monthly payments) and $Y = 30$ years, the monthly payment for Loan 2 is

$$PMT = \frac{P \times \left(\frac{APR}{n}\right)}{1 - \left(1 + \frac{APR}{n}\right)^{-nY}} = \frac{\$80,000 \times \left(\frac{0.075}{12}\right)}{1 - \left(1 + \frac{0.075}{12}\right)^{-12 \times 30}} = \$559.37.$$

You end up paying a total of $\$559.37$/payment $\times$ 360 payments $= \$201,373.20$.

The extra cost for Loan 1 is $1200 for closing costs. The extra cost for Loan 2 is $1200 for closing costs *plus* 1.5%

of the principal (the points). Thus the extra cost of Loan 2 is

$$\$1200 + (0.015 \times \$80,000) = \$1200 + \$1200 = \$2400.$$

Thus the total payout for Loan 1 over 30 years is
$$\$211,323.60 + \$1200 = \$212,523.60.$$

The total payout for Loan 2 over 30 years is
$$\$201,373.20 + \$2400 = \$203,773.20.$$

Loan 2 with the lower interest rate, even with points, is the better option.

23. First compute the monthly payment for each loan.

Using the loan payment formula with $P = \$80,000$, $APR = 0.0725$, $n = 12$ (for monthly payments) and $Y = 30$ years, the monthly payment for Loan 1 is

$$PMT = \frac{P \times \left(\frac{APR}{n}\right)}{1 - \left(1 + \frac{APR}{n}\right)^{-nY}} = \frac{\$80,000 \times \left(\frac{0.0725}{12}\right)}{1 - \left(1 + \frac{0.0725}{12}\right)^{-12 \times 30}} = \$545.74.$$

You end up paying a total of $\$545.74$/payment $\times$ 360 payments $= \$196,466.40$.

Using the loan payment formula with $P = \$80,000$, $APR = 0.065$, $n = 12$ (for monthly payments) and $Y = 15$ years, the monthly payment for Loan 2 is

$$PMT = \frac{P \times \left(\frac{APR}{n}\right)}{1 - \left(1 + \frac{APR}{n}\right)^{-nY}} = \frac{\$80,000 \times \left(\frac{0.065}{12}\right)}{1 - \left(1 + \frac{0.065}{12}\right)^{-12 \times 15}} = \$696.89.$$

You end up paying a total of $\$696.89$/payment $\times$ 180 payments $= \$125,440.20$.

The extra cost for Loan 1 is $\$1200$ *plus* 1% of the principal (the points), or

$$\$1200 + (0.01 \times \$80,000) = \$1200 + \$800 = \$2000.$$

The extra cost for Loan 2 is $\$1200$ costs *plus* 3% of the principal (the points), or

$$\$1200 + (0.03 \times \$80,000) = \$1200 + \$2400 = \$3600.$$

Thus the total payout for Loan 1 over 30 years is
$$\$196,466.40 + \$2000 = \$198,466.40.$$

The total payout for Loan 2 over 15 years is
$$\$125,440.20 + \$3600 = \$129,040.20.$$

Loan 2 with the lower interest rate, even with points and closing costs, is the better option.

25. How Much House Can You Afford? We can start with the loan payment formula and solve for the principal, P. This is done by dividing both sides of the formula by everything that multiplies P, then using the fact that division is the same as multiplication by the reciprocal:

$$PMT = \frac{P \times \left(\frac{APR}{n}\right)}{1 - \left(1 + \frac{APR}{n}\right)^{-nY}} \Rightarrow \frac{PMT}{\frac{\left(\frac{APR}{n}\right)}{1 - \left(1 + \frac{APR}{n}\right)^{-nY}}} = P \Rightarrow$$

$$P = \frac{PMT \times \left(1 - \left(1 + \frac{APR}{n}\right)^{-nY}\right)}{\left(\frac{APR}{n}\right)}$$

Now we have a formula for the principal in terms of the monthly payment. Now substituting $PMT = \$500$, $APR = 0.09$, $n = 12$ (monthly compounding), and $Y = 30$, we find that the principal is

$$P = \frac{PMT \times \left(1 - \left(1 + \frac{APR}{n}\right)^{-nY}\right)}{\left(\frac{APR}{n}\right)} = \frac{\$500 \times \left(1 - \left(1 + \frac{0.09}{12}\right)^{-12 \times 30}\right)}{\left(\frac{0.09}{12}\right)}$$

$$= \$62,140.93.$$

With monthly payments of $\$500$, you could afford a loan of about $\$62,000$.

If you can make a 20% down payment, then you will pay 80% of the price of the house with a loan. This means that $0.8P$ is the amount that you can finance, or

$$0.8P = \$62,140.93.$$

Dividing this equation by 0.8 and solving for P, we see that you can afford a house that costs $\$77,676.16$.

Unit 5D Income Taxes

Overview

Another aspect of most peoples' financial lives is income tax. The mathematics of income taxes is not difficult. However there are a lot of concepts and terminology involved with the subject. This unit is designed to give you a fairly thorough survey of income taxes, enough for you to make sense of complex tax forms and make critical decisions.

The first step in determining the amount of income tax you owe is to calculate your **taxable income**. From your **gross income** (the total of all your income), you subtract deductions and exemptions to arrive at your taxable income. Having computed your taxable income, the next step is to determine the tax itself. The U.S. tax system uses tax brackets or **marginal tax rates**: the percentage of your taxable income that you pay in taxes depends on your tax bracket. It also depends on your **filing status**, whether you are filing as a single person, a married person, or a head of household. This is also an appropriate place to discuss the **marriage penalty**.

Two other forms of income tax are **social security taxes** and **Medicare taxes** which apply to income from wages (not investments) and self-employment. We discuss the rules that govern these taxes.

Of great political and personal interest are the taxes assessed on **capital gains**, profits made from selling property or stocks. We will examine the (new) laws for capital gains and give several examples that illustrate the laws. Also of practical value are the tax implications of **tax-deferred savings plans** that many people use for retirement pensions and the **mortgage interest tax deduction** which is a large tax benefit for people holding mortgages.

In this unit you will find some relief from the big formulae and long calculations of previous units in this chapter. This chapter is very conceptual and it presents a lot of new terminology and practical information.

Key Words and Phrases

gross income	adjusted gross income	deductions
exemptions	dependents	standard deduction
itemized deduction	taxable income	tax credits
progressive income tax	marginal tax rates	filing status
social security tax	Medicare tax	capital gain
ordinary income	short-term capital gain	long-term capital gain
tax-deferred savings plan	mortgage interest deduction	marriage penalty

Key Concepts and Skills

- understand the basic terminology associated with income tax.
- compute taxable income from gross income given deductions and exemptions.
- use marginal tax rates to compute the tax on a given taxable income.
- compute the social security and Medicare taxes on a given income.
- understand and determine taxes on capital gains.
- incorporate tax-deferred savings plans and mortgage interest deductions into a tax calculation.

Solutions Unit 5D

Tax Calculations.

1. Suzanne's gross income is her wages plus her interest, or $33,200 + $350 = $33,550. Her contribution to a tax-deferred plan is an adjustment to her gross income. So her AGI (adjusted gross income) is $33,550 − $500 = $33,050. She can take a personal exemption of $2650 and the standard deduction of $4150 (because her itemized deductions do not exceed the standard deduction). This gives her a taxable income of

$$\$33,050 - \$2650 - \$4150 = \$26,250.$$

3. Wanda's gross income consists of salary and interest which total $35,400 + $500 = $35,900. With no adjustments, her AGI (adjusted gross income) is also $35,900. Her total exemptions are 3 × $2650 = $7950 and she should take the standard deduction of $3450. So her taxable income is

$$\$35,900 - \$7950 - \$3450 = \$24,500.$$

Tax Credits and Tax Deductions.

5. Karen and Tremaine will save a full $500 in taxes because of their tax credit.

7. Lisa will have no tax savings because she claims the standard deduction, so her $1000 charitable contribution is not used as a deduction.

9. Sebastian's $1000 charitable contribution reduces his taxable income by $1000, so he will save 28% of $1000 or $280.

Marginal Tax Calculations.

11. Gene is single and in the 28% marginal tax bracket, so his first $24,650 is taxed at 15% and the remaining income is taxed at 28%. His tax bill is

$$15\% \times \$24,650 + 28\% \times (\$35,400 - \$24,650) = \$6708.$$

13. Bobbi is in the 36% marginal tax bracket for married people filing separately. Her first $20,600 is taxed at 15%, her income between $20,601 and $49,800 is taxed at 28%, her income between $49,801 and $75,875 is taxed at 31%, and the remaining income is taxed at 36%. Her tax bill is

$$15\% \times \$20,600 + 28\% \times (\$49,800 - \$20,600) + 31\% \times$$
$$(\$75,875 - \$49,800) + 36\% \times (\$77,300 - \$75,875) =$$
$$\$19,862.$$

15. Paul is in the 31% marginal tax bracket for head of household. His first $33,050 is taxed at 15%, his income between $33,051 and $85,350 is taxed at 28%, and the remaining income is taxed at 31%. His tax bill is

$$15\% \times \$33,050 + 28\% \times (\$85,350 - \$33,050) + 31\% \times$$
$$(\$89,300 - 85,350) = \$20,826.$$

With the $500 tax credit his tax bill is reduced by $500, so his final tax bill is $20,326.

17. This couple is in the 31% marginal tax bracket for married people filing jointly. Their first $41,200 is taxed at 15%, their income between $41,201 and $99,600 is taxed at 28%, and the remaining income is taxed at 31%. Their tax bill is

$$15\% \times \$41,200 + 28\% \times (\$99,600 - \$41,200) + 31\% \times$$
$$(\$105,500 - 99,600) = \$24,360.$$

With a $1000 tax credit, their tax bill is reduced to $23,360.

19. Marriage Penalty.

a. If Joan and Peter file as a married couple, their combined adjusted gross income is $44,500 + $33,400 = $77,900. They can claim two exemptions for a reduction of 2 × $2650 = $5300. They would take the standard deduction for married couples of $6900. So their taxable income is

$$\$77,900 - \$5300 - \$6900 = \$65,700.$$

This places them in the 28% marginal tax bracket. Their tax bill is

$$15\% \times \$41,200 + 28\% \times (\$65,700 - \$41,200) = \$13,040.$$

b. Let's see what happens if Joan and Peter file separately as individuals. Joan's taxable income is

$$\$44,500 - \$2650 - \$4150 = \$37,700,$$

(subtracting one exemption and a standard deduction). This puts her in the 28% marginal tax bracket with a tax bill of

$$15\% \times \$24,650 + 28\% \times (\$37,700 - \$24,650) = \$7352.$$

Similarly, Peter's taxable income is

$$\$33,400 - \$2650 - \$4150 = \$26,600.$$

This also puts him in the 28% marginal tax bracket with a tax bill of

$$15\% \times \$24,650 + 28\% \times (\$26,600 - \$24,650) = \$4244.$$

Their combined tax bill filing as two individuals is $7352 + $4244 = $11,596, which is less than their tax bill as a married couple.

FICA Taxes.

21. Lars' entire income of $28,000 is subject to the 7.65% FICA tax:

$$\text{FICA tax: } 7.65\% \times \$28,000 = \$2142.$$

To find his federal income tax, note that his adjusted gross income is $28,000 − $2500 = $25,500, due to the tax-deferred contribution. We then subtract the $2650 personal exemption and $4150 standard deduction to get his taxable income:

$$\text{taxable income} = \$25,500 - \$2650 - \$4150 = \$18,700$$

This income is taxed at the 15% marginal rate:

$$\text{federal tax} = 15\% \times \$18,700 = \$2805.$$

Thus Lars' FICA tax and income tax total is $2142 + $2805 = $4947 which is

$$\frac{\$4947}{\$28,000} = 0.170 = 17.7\%$$

of his gross income. Thus his overall tax rate is 17.7%.

23. Jack's salary of $44,800 is subject to the 7.65% FICA tax:

FICA tax: 7.65% × $44,800 = $3427.

To find his income tax, note that his gross income is salary plus interest:

gross income = $44,800 + $1250 = $46,050

Therefore his adjusted gross income is $46,050 − $2000 = $44,050, due to his tax-deferred contribution. We then subtract the $2650 personal exemption and $4150 standard deduction to get her taxable income:

taxable income = $44,050 − $2650 − $4150 = $37,250

The first $24,650 of this income is taxed at the 15% marginal rate, the rest at the 28% rate:

federal tax = 15% × $24,650 + 28% × ($37,250 − $24,650) = $7226.

Thus Jack's FICA tax and income tax total is $3427 + $7226 = $10,653 which is

$$\frac{\$10,653}{\$46,050} = 0.231 = 23.1\%$$

of his gross income. Thus his overall tax rate is 23.1%.

25. Because she is self-employed, Brittany's salary of $48,200 is subject to the 15.3% FICA tax:

FICA tax: 15.3% × $48,200 = $7375

Now recall the technicality that for self-employed people, half of the FICA tax (in this case, $3687) can be taken as an adjustment to the gross income. Therefore Brittany's adjusted gross income is $48,200 − $3687= $44,513. We then subtract the $2650 personal exemption and $4150 standard deduction to get her taxable income:

taxable income = $44,513 − $2650 − $4150 = $37,713

This puts Brittany in the 28% marginal tax bracket as a single person. Her federal tax is

federal tax = 15% × $24,650 + 28% × ($37,713 − $24,650) = $7355.

Thus Brittany's FICA tax and income tax total is $7375 + $7355 = $14,730 which is

$$\frac{\$14,730}{\$48,200} = 0.306 = 30.6\%$$

of her gross income. Thus her overall tax rate is 30.6%.

Capital Gains vs. Ordinary Income.

27. Pierre has only ordinary income and no capital gains; his gross income and adjusted gross income are $120,000, all of which is subject to FICA tax and income tax. For the FICA tax, the first $65,000 is taxed at a rate of 7.65%, the rest is taxed at 1.45%. Thus his FICA tax is

7.65% × $65,000 + 1.45% × (120,000 − $65,000) = $5770.

To find his income tax, we can subtract the $4150 standard deduction (the personal exemption) to get Pierre's taxable income:

taxable income = $120,000 − $4150 = $115,850

This puts Pierre in the 31% marginal tax bracket as a single person. His income tax is

income tax = 15% × $24,650 + 28% × ($59,750 − $24,651) + 31% × ($115,850 − $59,750) = $30,917.

Thus Pierre's FICA tax and income tax total is $5770 + $30,917 = $36,687 which is

$$\frac{\$36,687}{\$120,000} = 0.306 = 30.6\%$$

of his gross income. Thus his overall tax rate is 30.6%.

By contrast, all of Katherine's income is capital gains. She owes no FICA taxes. Her gross income can be reduced by the $4150 standard deduction to give her taxable income (we were told to neglect personal exemptions):

taxable income = $120,000 − $4150 = $115,850

As capital gains for a single person, this income is taxed at 15% for the first $24,650 and the remainder at 20%. So Katharine's income tax is

15% × $24,650 + 20% × ($115,850 − $24,650) = $21,938.

Thus Katharine's FICA tax and income tax total is $0 + $21,937 = $21,938 which is

$$\frac{\$21,938}{\$120,000} = 0.183 = 18.3\%$$

of her gross income. Thus her overall tax rate is 18.3%. Although Pierre and Katharine have the same gross income, Katharine pays considerably less in taxes because her income consists of capital gains.

29. Because he is self-employed, Fred's salary of $275,000 is subject to a FICA tax of 15.3% on the first $65,000 and 2.9% of the remainder. Thus his FICA tax is

15.3% × $65,000 + 2.9% × ($275,000 − $65,000) = $16,035.

To compute Fred's income tax, recall the technicality that for self-employed people, half of the FICA tax (in this case, $8017) can be taken as an adjustment to the gross income. Therefore Fred's adjusted gross income is $275,000 − $8017 = $266,983. He can also subtract the $4150 standard deduction (neglecting the personal exemption) to get his taxable income:

taxable income = $266,983 − $4150 = $262,833

This puts Fred in the 36% marginal tax bracket as a single person. His income tax is

income tax = 15% × $24,650 + 28% × ($59,750 − $24,650) + 31% × ($124,650 − $59,750) + 36% × (262,833 − $124,650) = $83,390.

Thus Fred's FICA tax and income tax total is $16,035 + $83,390 = $99,425 which is

$$\frac{\$99,425}{\$275,000} = 0.362 = 36.2\%$$

of his gross income. Thus his overall tax rate is 36.2%. By contrast, Tamara has all of her $275,000 income in capital gains. She owes no FICA taxes. Her gross income can be reduced by the $4150 standard deduction to give her taxable income:

taxable income = $275,000 − $4150 = $270,850

As capital gains for a single person, this income is taxed at 15% for the first $24,650 and the remainder at 20%. So Tamara's income tax is

15% × $24,650 + 20% × ($270,850 − $24,650) = $52,938.

Thus Tamara's FICA tax and income tax total is $0 + $52,938= $52,938 which is

$$\frac{\$52,938}{\$275,000} = 0.193 = 19.3\%$$

of her gross income. Thus her overall tax rate is 19.2%. Although Fred and Tamara have the same gross income, Tamara pays considerably less in taxes because her income consists of capital gains.

31. With a taxable income of $45,000, your marginal tax rate is 28%. Thus each $600 contribution to a tax-deferred savings plan will reduce your tax bill by

28% × $600 = $168.

In other words, $600 will go into your tax-deferred savings account each month, but your monthly paychecks will decrease by only

$600 − $168 = $432.

33. With a taxable income of $150,000, your marginal tax rate is 36%. Thus each $800 contribution to a tax-deferred savings plan will reduce your tax bill by

36% × $800 = $288.

In other words, $800 will go into your tax-deferred savings account each month, but your monthly paychecks will decrease by only

$800 − $288 = $512.

35. Mortgage Tax Savings. With a $120,000 mortgage and an *APR* of 8%, the annual interest payments will be very close to 8% × $120,000 =$9600 during the early years of the loan. A taxable income of $55,000 puts you in the 28% marginal tax bracket. This $9600 of interest is tax deductible; because you are single, it is $5450 larger than the $4150 standard deduction. Thus the additional $5450 deduction saves you

28% × $5450 = $1526

in taxes over the year — which is a monthly savings of about $127.

37. Home Equity Loan.

a. The first few payments of your credit card debt will be mostly interest. The monthly interest rate is 18%/12 = 1.5%. So your interest in the early months is about 1.5% × $4000 = $60.

b. The first few payments of your home equity loan will also be mostly interest. The monthly interest rate is 10%/12 = 0.83%. So your interest in the early months is about 0.83% × $4000 = $33.

c. Thus the difference in interest payments between the two loans is $60 − $33 = $27. Because you are in the 28% marginal tax bracket and the interest on the home equity loan is deductible, you save 28% × $33 = $9.24 per month. Thus, the effective interest on the home equity loan is $33 − $9.24 = $23.76. It makes good sense to convert your credit card debt to a home equity loan. Not only are the interest payments less, but they produce savings because they are tax deductible. Of course, you must have equity in a house!

Unit 5E Investments

Overview

In this closing unit of the finance chapter, we look into forms of investments other than bank accounts. Some key considerations in choosing investment plans are **liquidity** (how easily your money can be withdrawn), **risk** (how safe your money is), and **return** (the amount that you earn on your investment). Liquidity is fairly easy to determine, risk is less easy to assess, and for return there is a fairly simple formula that gives the annual return on an investment.

We examine four general types of investments in light of these three factors:

- small company stocks,
- large company stocks,
- corporate bonds, and
- Treasury bills.

These four types of investment have very different characteristics in terms of liquidity, risk, and return.

An essential part of investing money is keeping track of how well your investments are doing. Many people follow their investments using the financial pages of the newspapers. For this reason, we spend some time discussing how to read the financial pages for stocks, bonds, and mutual funds. Bonds in particular have some special terminology (such as **face value**, **coupon rate**, and **discount**) and a special formula for calculating the **current yield** on a bond.

Key Words and Phrases

stocks	bonds	cash
mutual fund	liquidity	risk
return	total return	annual returns
portfolio	corporation	shares
market price	dividends	PE ratio
percent yield	face value	coupon rate
maturity date	discount	points
load	current yield	annual fee

Key Concepts and Skills

- understand the terms liquidity, risk, and return as they apply to various forms of investment.
- determine the annual yield on an investment given the total return.
- understand the terms used to report stocks in the financial pages of a newspaper.
- understand the terms used to report bonds in the financial pages of a newspaper.
- understand the terms used to report mutual funds in the financial pages of a newspaper.
- compute the current yield of a bond.

Solutions Unit 5E

1. Liquidity, Risk, and Return.

a. The 30-year term of the bond means that the bond is not liquid at all; unless you sell the bond to someone, it will be 30 years before you can get your principal out. The U.S. Treasury bond is a very safe investment. The annual return is fixed at 6%.

b. Large company stocks are fairly liquid because they are easy to sell. All stocks are inherently risky, but stock in a large established company *usually* carries a relatively low risk because it's unlikely that such a company will go out of business. If the company does well, you can expect a moderate return on your investment. Of course, a dramatic increase in the stock value is also unlikely.

c. A small company stock may be more difficult to sell than a large company stock, making it somewhat less liquid. Such a stock also is extremely risky: if the company's software is not successful, the company and its stock may be worthless. On the other hand, if the company succeeds, the company may flourish and provide a huge return on your investment.

d. The 15% annual interest rate represents a high return compared to U.S. Treasury bonds, but the company's poor financial health also makes it very risky: If the company goes bankrupt and does not recover, you'll never get your money back. This 5-year bond also has poor liquidity because the company is not obligated to pay you back for 5 years and the bond's risky nature may make it difficult to sell in the meantime.

3. Annual Returns.
We will need the annual return formula for these problems. Recall that if the total return on an investment is known then the annual return is given by the formula:

$$\text{annual return} = (\text{total return} + 1)^{1/Y} - 1,$$

a. The cost of the stock was 100 shares × $55/share = $5500. The sale of the stock was worth $10,300. This means the total return was

$$\frac{\text{new value} - \text{previous value}}{\text{previous value}} = \frac{\$10,300 - \$5500}{\$5500} = 0.873 = 87.3\%$$

This total return results in an annual return after $Y = 20$ years of

$$(\text{total return} + 1)^{1/Y} - 1 = (0.873 + 1)^{1/5} - 1 = 1.873^{0.2} - 1 = 0.134 = 13.4\%.$$

b. The cost of the bond was $8000. The sale price of the bond was $12,500. This means the total return was

$$\frac{\text{new value} - \text{previous value}}{\text{previous value}} = \frac{\$12,500 - \$8000}{\$8000} = 0.563 = 56.3\%$$

This total return results in an annual return after $Y = 5$ years of

$$(\text{total return} + 1)^{1/Y} - 1 = (0.563 + 1)^{1/20} - 1 = 1.563^{0.05} - 1 = 0.023 = 2.3\%.$$

c. The mutual fund was bought for $5500 and sold for $11,300. This means the total return was

$$\frac{\text{new value} - \text{previous value}}{\text{previous value}} = \frac{\$11,300 - \$5500}{\$5500} = 1.055 = 105.5\%$$

This total return results in an annual return after $Y = 10$ years of

$$(\text{total return} + 1)^{1/Y} - 1 = (1.055 + 1)^{1/10} - 1 = 2.055^{0.1} - 1 = 0.075 = 7.5\%.$$

d. The stock was bought for $4500 and sold for $2500 (at a loss). This means the total return was

$$\frac{\text{new value} - \text{previous value}}{\text{previous value}} = \frac{\$2500 - \$4500}{\$4500} = -0.444 = -44.4\%$$

This total return results in an annual return after $Y = 3$ years of

$$(\text{total return} + 1)^{1/Y} - 1 = (-.444 + 1)^{1/3} - 1 = -0.178 = -17.8\%.$$

5. Historical Returns.
Your great uncle's investments of $500 were made 70 years ago. At the annual returns given in Table 5.3, the current value of the investments are given by the compound interest formula:

Small Stocks (Annual return 12.5%)

$$\text{Current value} = \$500 \times (1 + 0.125)^{70} = \$1,903,911.$$

Large Stocks (Annual return 10.5 %

$$\text{Current value} = \$500 \times (1 + 0.105)^{70} = \$542,412.$$

Bonds (Annual return 5.7 %)

$$\text{Current value} = \$500 \times (1 + 0.057)^{70} = \$24,223.$$

Treasury Bills (Annual return 3.7 %)

$$\text{Current value} = \$500 \times (1 + 0.037)^{70} = \$6360.$$

Reading Stock Tables.

7. Refer to Figure 5.5 and look at the line for the Mossimo.

a. The ticker symbol is MGX.

b. The high and low prices yesterday were $8.63 and $8.38, respectively.

c. The closing price yesterday was $8.38. The net change in price from two days ago was −$0.13, so the stock dropped in price by $0.13. This means the stock closed at $8.51 two days ago.

d. The number of shares that traded yesterday was $100 \times 100 = 10,000$.

e. This stock does not pay dividends.

f. Because PE ratio = (share price)/(earnings per share), we also know that earnings per share = (share price)/(PE ratio) (see Example 6 of the unit). With a PE ratio of 20 and a share price of $8.38, this stock had earnings per share of $8.38/20 = $0.42. The current share price is 20 times the earnings per share.

13. Total Return on Stock. The closing price for Motorola in Figure 5.5 is $71.38. If you paid $72 per share, then you lost $72.00 − $71.38 = $0.62 per share in the transaction. With the $0.48 per share dividend, you lost $0.14 per share. If you also paid $1 per share as a fee, then you lost $1.14 per share.

15. Total Return on Stock. The closing price for Mossimo in Figure 5.5 is $8.38. If you paid $46.00 per share, then you lost $46.00 − $8.38 = $37.62 per share in the transaction. If you also paid $0.25 per share as a fee, then you lost $37.87 per share.

17. Bond Yields.

a. The interest payment on this bond is $6\% \times \$10,000 = \$600$. The market value is $9500. Therefore, the current yield is

$$\text{current yield} = \frac{\text{annual interest payment}}{\text{market price of bond}} = \frac{\$600}{\$9500} = 6.3\%$$

b. The interest payment on this bond is $7\% \times \$10,000 = \$700$. The market value is $10,500. Therefore, the current yield is

$$\text{current yield} = \frac{\text{annual interest payment}}{\text{market price of bond}} = \frac{\$700}{\$10,500} = 6.7\%$$

c. The interest payment on this bond is $8\% \times \$10,000 = \$800$. The market value is $9000. Therefore, the current yield is

$$\text{current yield} = \frac{\text{annual interest payment}}{\text{market price of bond}} = \frac{\$800}{\$9000} = 8.9\%$$

19. Bond Interest.

a. Because this bond is quoted at 105 points, its market value is $1.05 \times \$10,000 = \$10,500$. Using the current yield formula with a yield of 8.5%, we see that

$$\text{current yield} = \frac{\text{annual interest payment}}{\text{market price of bond}} \Rightarrow 8.5\% = \frac{\text{interest}}{\$10,500}.$$

Multiplying both sides of this expression by $10,500, we find that the annual interest is

$$\text{interest} = \$10,500 \times 0.085 = \$892.50.$$

b. Because this bond is quoted at 98 points, its market value is $0.98 \times \$10,000 = \$9800$. Using the current yield formula with a yield of 6.5%, we see that

$$\text{current yield} = \frac{\text{annual interest payment}}{\text{market price of bond}} \Rightarrow 6.5\% = \frac{\text{interest}}{\$9800}.$$

Multiplying both sides of this expression by $9800, we find that the annual interest is

$$\text{interest} = \$9800 \times 0.065 = \$637.00.$$

c. Because this bond is quoted at 102.5 points, its market value is $1.025 \times \$10,000 = \$10,250$. Using the current yield formula with a yield of 7.0%, we see that

$$\text{current yield} = \frac{\text{annual interest payment}}{\text{market price of bond}} \Rightarrow 7.0\% = \frac{\text{interest}}{\$10,250}.$$

Multiplying both sides of this expression by $10,250, we find that the annual interest is

$$\text{interest} = \$10,250 \times 0.07 = \$717.50.$$

21. Mutual Fund Growth. According to Figure 5.7, the Calvert Income Fund had a total return (over the last three years) of 8.0%. This means that over three-years the account increased by a factor of 1.08. If you invested $500 in the fund three years ago, it would now be worth $500 \times 1.08 = \$540$.

6. MODELING OUR WORLD

Overview

Just as a map or a globe is a model of the Earth and a set of floor plans is a model of a building, we can also create mathematical models that represent real-world problems and situations. In fact, it might be argued that one of the fundamental goals of mathematics is to create mathematical models.

In this chapter, we introduce perhaps the most basic tool of mathematical modeling: functions. The presentation of functions in this book may be different than those you have seen elsewhere. We start at the beginning and develop the idea of a function in a practical and visual way. Hopefully, in this way, whether you are seeing functions for the first time or not, you will become comfortable with this essential mathematical concept.

Unit 6A Functions: The Building Blocks of Mathematics Models

Overview

This unit is brief and has a single purpose. Using words, a few definitions, pictures, and tables, the goal is to give a very qualitative idea of a **function**. There is no mysterious notation, no x's and y's, just an intuitive introduction to functions.

A function is a relationship between two quantities or **variables**. A function could describe how your height increases in time or how the temperature decreases with altitude. We call the variables the **independent variable** and the **dependent variable**. The terminology arises because we usually think of the dependent variable *changing with respect to* the independent variable; that is, if we make a change in the independent variable, it produces a change in the dependent variable.

There are four different ways to visualize or represent a function.

- a data table of values of the two variables,
- in words,
- with a graph (picture), and
- using an equation or formula.

In the remainder of this unit, we explore the first three ways of representing functions. The first two approaches (a data table and in words) are straightforward and probably familiar. So we focus on graphing functions. The use of equations will be studied in the next unit.

Two concepts are important in graphing functions.

- The **domain** of a function is the set of values that both make sense and are of interest for the *independent variable*.
- The **range** of a function consists of the values of the *dependent variable* that correspond to the values in the domain.

If you can identify the domain and range of a function, then you have saved yourself a lot of work. You need to make the graph only for those values of the independent variable in the domain and for those values of the dependent variable in the range. Having defined the domain and range, we present several examples of functions and their graphs. The highlight box on creating and using graphs summarizes the graphing process.

Key Words and Phrases

mathematical model	function	variable
independent variable	dependent variable	graph
coordinate plane	axis	origin
coordinates	quadrants	domain
range		

Key Concepts and Skills

- find the domain and range of a function given in table form or word form.
- use the domain and range of a function to scale the axes of a graph of the function.
- know when it makes sense to "fill in" between the points of a graph
- use the four-step process to create a graph of a function given in the form of a data table.
- use the four-step process to create a rough graph of a function given in a descriptive (word) form.

Solutions Unit 6A

Related Quantities.

3.

a. As the weight of the bag of apples increases, the price of the apples also increases.

b. Movies in 1960 cost about $1. Movies in 1998 cost about $7. The price of movies increases as time goes on.

c. As the price of a product increases, the demand (the number of items that can be sold) generally decreases.

d. According to Newton's Law of Gravitation, the strength of the Earth's gravitational force decreases as the distance from the Earth increases.

5.

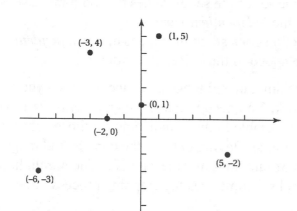

Functions from Graphs.

7.

a. The independent variable is time, measured in years, and the dependent variable is world population.

b. Because of the data values given, we can take the domain to be the years between 1950 and 1990. The range is all populations between about 2.5 billion and 6 billion.

c. The function shows a steadily increasing world population between 1950 and 1990 (and beyond).

Functions from Data Tables.

9. (a) The variables are (*time, temperature*) or (*date, temperature*). (b) The domain is all days over the course of a year. The range is temperatures between 38° and 85° (although for graphing purposes the range could be made larger). (c) The graph of the data is shown below.

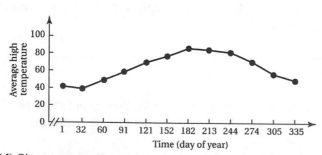

(d) Since temperature varies smoothly throughout the year, it would make sense to fill in the graph between the data points. The resulting smoothed graph would show variations in temperature over the course of a year, but would not show day-to-day variations very well.

11. (a) The variables are (*time, projected population*). (b) The domain is all years between 1995 and 2030. The range is population values between 258 million and 302

million, although for graphing purposes, it might be better to use a larger interval such as 250 million to 310 million. (c) The graph of the data is shown below.

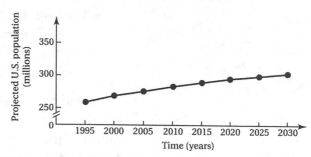

(d) Because populations vary smoothly in time, it would make sense to fill in the graph between the data points. The resulting smoothed graph would show long-term variations, but would not show day-to-day variations.

13. (*Altitude, Pressure*) **Function.**

a. Referring to Figure 6.5 and reading the value of pressure (in inches of mercury) at the specified altitudes gives the approximate pressure values (8000 ft, 23 in), (17,000 ft, 16 in), and (25,000 ft, 12 in).

b. Referring to Figure 6.5, the altitudes at which the pressure has the values 27, 20, and 12 inches of mercury are 5000 ft, 12,000 ft, and 25,000 ft.

c. Extending the graph of Figure 6.5, it appears that the pressure reaches 5 inches of mercury at an altitude of about 50,000 ft.

d. Atmospheric pressure varies from day to day and from one location to another. If accurate information is needed, a specific pressure graph should be used.

Rough Sketches of Functions.

7. Example: The domain of the function (*altitude, temperature*) is all altitudes of interest, say, 0 ft to 15,000 ft (or 0 meters to 4000 meters). The range is all temperatures associated with the altitudes in the domain; the interval 30°F to 90°F (or about 0°C to 30°C) would cover all temperatures of interest. A rough sketch of this decreasing function is shown below. With some reliable data, this graph is a good model of how the temperature varies with altitude.

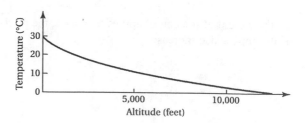

8. The domain of the (*blood alcohol level, reflex time*) function consists of all possible blood alcohol levels expressed as a percentage; for example, numbers between 0% and 0.25% would be appropriate. The range would consist of the reflex times associated with those blood alcohol levels. A rough sketch of this decreasing function is shown below. The validity of this graph as a model of alcohol impairment will depend on how accurately reflex times can be measured.

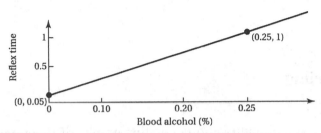

9. The domain of the function (*mortgage interest rate, homes sold*) consists of typical interest rates, 6% to 12%. The range is a little difficult to define precisely; it consists of the number of homes sold per year in a given town for the various interest rates. At low interest rates, the number of homes sold would be relatively high. This number would decrease with increasing interest rates. The graph of this function may not be a very good model unless reliable data could be found. Generally speaking, interest rates do not remain fixed for an entire year.

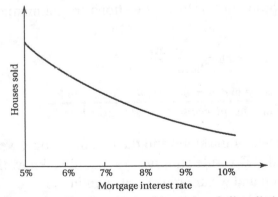

10. Example: The domain would consist of all realistic gasoline prices, say between $1 per gallon and $3 per gallon. The range includes all possible or reasonable numbers of tourists in Yellowstone, probably given per day or per week. The graph below shows the general trend that as gasoline prices increase, tourists are less likely to travel and so the number of tourists decreases. The graph provides a qualitative model of this relation.

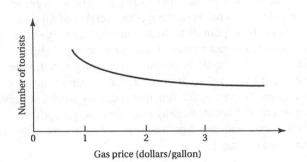

11. Example: The domain of the relation (*minutes after lighting, length of candle*) consists of all times between 0 minutes and the expected time for the candle to disappear, say, 0 minutes to 100 minutes. The range is the set of numbers between zero and the initial length of the candle, say, 0 inches to 5 inches. A rough sketch of this decreasing relation is shown below. With regular measurement of candle length, this graph will be a good model.

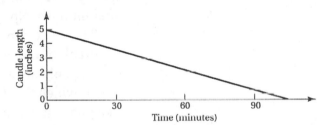

12. The domain of the function (*time of day, elevation of tide*) consists all times over a two-day period; it could be all numbers between 0 hours and 48 hours, or all times between, say, 6:00 AM on Friday and 6:00 AM on Sunday. The range consists of the heights of the tide over that two-day period. Some research would be needed to find the exact tide heights at a given location. But the graph would be periodic, rising and falling twice a day. The graph would be a very good model provided it is based on accurate data.

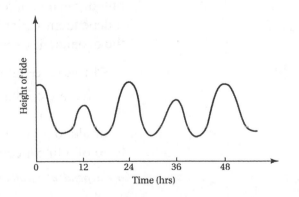

13. The domain of the function (*weight of car, average gas mileage*) consists of the typical weights of all cars, say from 1000 pounds to 4000 pounds. The range consists of the average gas mileage for cars of various weights. This function should be roughly decreasing because we expect gas mileage to decrease as the weight of the car increases. However, the function probably would not give a good model because many cars could be exceptions (heavy cars with good gas mileage and light cars with poor gas mileage).

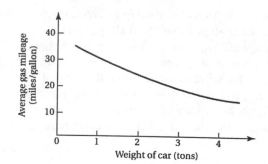

Unit 6B Linear Modeling

Overview

From the previous unit, you probably appreciate that graphs of functions can take many different forms. In this unit we study one very special, but widely used, family of functions — those functions whose graphs are straight lines. Not surprisingly, these functions are called **linear functions**.

The most important property of a linear function is its **rate of change**. If the graph of two variables is a straight line, it means that a fixed change in one variable always produces the *same* change in the other variable. The rate at which the dependent variable changes with respect to the independent variable is called the rate of change of the function. For linear functions, the rate of change is constant and it is equal to the **slope** of the graph. These two fundamental properties of linear functions are summarized in the following rules.

$$\text{slope of a linear graph} = \frac{\text{vertical } \textit{rise}}{\text{horizontal } \textit{run}}$$

$$\text{rate of change} = \text{slope} = \frac{\text{change in } \textit{dependent variable} \text{ from } P_1 \text{ to } P_2}{\text{change in } \textit{independent variable} \text{ from } P_1 \text{ to } P_2}.$$

Having established the equivalence of the slope and the rate of change, we introduce the **rate of change rule**. This rule simply says that if we know the rate of change of a linear function and we are given a change in the independent variable, then we can determine the corresponding change in the dependent variable:

Change in dependent variable = (rate of change) × (change in independent variable).

The rate of change rule is really just a stepping stone to the final goal of the unit — to write a general equation for a linear function. After a detailed example to motivate the idea of a linear equation, we present the general form of a linear equation. It looks like this:

dependent variable = initial value + (rate of change × *independent variable*)

Up to this point we have used words for the variable names. But you will soon see that this practice gets cumbersome. So for the sake of economy (not confusion!) we start to use single letter names for variables. For example, instead of writing *time*, we just use t; and instead of writing *number of chips*, we just use N. Don't let the use of letters confuse you; it just makes working with linear equations easier.

After several examples of creating and using linear equations, we come to one last topic. A linear equation it is a model or a compact description of a particular situation. Once we have it, it can be used for prediction or to answer other useful questions. We want to be able to answer questions such as:

- when the independent variable has a certain value, what is the corresponding value of the dependent variable?
- when the dependent variable has a certain value, what is the corresponding value of the independent variable?

This brings us to the necessity of *solving* linear equations when we are given a particular value of either the dependent variable or the independent variable. Two simple rules are all we need to solve any linear equation for either variable:

- we can always add or subtract the same quantity from both sides of an equation.
- We can always multiply or divide both sides of an equation by a (nonzero) quantity.

Several examples conclude this action-packed unit. There is a lot of material in this unit and it is best to work at it slowly and give yourself plenty of time for reading and for practice.

Key Words and Phrases

linear function	linear graph	rate of change
slope	rate of change rule	initial value
general linear equation		

Key Concepts and Skills

- determine the slope of a straight line given two points on the line.
- determine the rate of change of a linear function, either from a description of the function or from two points associated with the function.

- use the rate of change rule to compute the change in the dependent variable given the rate of change and a change in the independent variable.
- find the equation of a linear function given the rate of change and the initial value.
- evaluate a linear function for the dependent variable given a value of the independent variable.
- solve a linear function for the independent variable given a value of the dependent variable.
- create a linear function from information about its rate of change and initial value.

Solutions Unit 6B

1. Rates of Change.

a. To find the slope (rate of change) of this function, we need to identify two points on the graph. Clearly the point (0, 0) is on the graph. Another point is (1, 0.75) Using the formula for the slope of a straight line graph, we have

$$\text{slope} = \frac{\text{change in dependent variable}}{\text{change in independent variable}} = \frac{(0.75 - 0) \text{ in}}{(1 - 0) \text{ hr}} = 0.75 \frac{\text{in}}{\text{hr}}.$$

The rate of change for this storm is 0.75 inches per hour.

b. To find the slope (rate of change) of this function, we can identify the two points on the graph (0, 0) and (1, 1.25) Using the formula for the slope of a straight line graph, we have

$$\text{slope} = \frac{\text{change in dependent variable}}{\text{change in independent variable}} = \frac{(1.25 - 0) \text{ in}}{(1 - 0) \text{ hr}} = 1.25 \frac{\text{in}}{\text{hr}}.$$

The rate of change for this storm is 1.25 inches per hour.

Analyzing Linear Graphs.

3.

a. The graph gives a picture of a trip during which some travelers started 500 miles from home and drove at a constant speed toward home, arriving eight hours after leaving. The change in the dependent variable (*distance*) is −500 miles and the change in the independent variable (*time*) is 8 hours. So the slope (which corresponds to the speed of travel) is

$$(-500 \text{ miles})/(8 \text{ hours}) = -62.5 \text{ mi/hr}.$$

The negative sign simply means that the distance from home decreases during the trip. If the speed really is constant throughout the trip, this graph is a realistic model.

b. This graph expresses the function between the price of a product and the number of items that can be sold at that price (the demand). Not surprisingly, as the cost increases, the demand decreases. Two points on the straight line are ($0, 1000 units) and ($36, 0 units). The slope of the line is

(0 units − 1000 units)/($36 − $0) = −27.8 units/dollar

which means that for every $1 that the price increases, the demand decreases by 27.8 units. The demand function does decrease with price in practice, although it is not necessarily a linear function.

c. This graph expresses the familiar function between the Celsius and Fahrenheit temperature scales. Two points on the straight line are (32°F, 0°C) and (0°F, −18°C), so the slope is

(−18°C − 0°C)/(0°F − 32°F) = 0.56 °C/°F.

This slope means that a change of 1°F is equal to a change of 0.56 °C = 5/9°C. This linear graph is an exact model for the conversion between the temperature scales.

d. The graph says that gas mileage increases as driving speed increases. Two points on the line are (30 mi/hr, 20 mi/gal) and (70 mi/hr, 30 mi/gal). The slope of the line is

(30 mi/gal − 20 mi/gal)/(70 mi/hr − 30 mi/hr) = 0.25 (mi/gal)/(mi/hr)

which means that for every mi/hr increase in driving speed the gas mileage increases by 0.25 mi/gal. This linear function is not realistic; gas mileage has a maximum around 35 mi/hr and is lower for both high speeds and low speeds.

Rate of Change Rule. In all of these problems the rate of change rule can be used:

Change in dependent variable = rate of change × change in independent variable.

5. The water depth decreases with respect to time at a rate of 0.25 inches per hours. The rate of change is −0.25 in./hr. In 6.5 hours, the water depth changes by

$$-0.25 \text{ in./hr} \times 6.5 \text{ hr} = -1.625 \text{ in,}$$

a decrease of 1.625 inches. In 12.5 hours, the water depth changes by

$$-0.25 \text{ in./hr} \times 12.5 \text{ hr} = -3.125 \text{ in,}$$

a decrease of 3.125 inches.

7. The tree diameter increases by 0.2 inches per year. The rate of change is 0.2 in./yr. In 4.5 years, the tree will increase in diameter by

$$0.2 \text{ in./yr} \times 4.5 \text{ years} = 0.9 \text{ inches.}$$

In 20.5 years, the tree will increase in diameter by

$$0.2 \text{ in./yr} \times 20.5 \text{ years} = 4.1 \text{ inches.}$$

9. The candle decreases in length at 2 centimeter per hour. The rate of change is −2 cm/hr. In 3.8 hours the length of the candle changes by

$$-2 \text{ cm/hr} \times 3.8 \text{ hr} = -7.6 \text{ cm,}$$

a decrease of 7.6 centimeters. In 4.2 hours the length of the candle changes by

$$-2 \text{ cm/hr} \times 4.2 \text{ hr} = -8.4. \text{ cm,}$$

a decrease of 8.4 centimeters.

11. The snow depth increases by 4 inches per hour. The rate of change is 4 in./hr. In 5.5 hours the snow depth will increase by

$$4 \text{ in./hr} \times 5.5 \text{ hr} = 22 \text{ in.}$$

In 7.8 hours the snow depth will increase by

$$4 \text{ in./hr} \times 7.8 \text{ hr} = 31.2 \text{ in.}$$

13. The boiling point of water at sea level is 212 °F. The boiling point decreases by 2°F for every increase of 1000 feet of elevation. The rate of change is −2°F/1000 ft = −0.002 °F/ft. At an elevation of 6000 ft above sea level, the change in boiling point is

$$-2°F/1000 \text{ ft} \times 6000 \text{ ft} = -12°F.$$

Thus the boiling point at 6000 feet is 212 °F − 12°F = 200 °F. At an elevation of 12,000 ft above sea level, the change in boiling point is

$$-2°F/1000 \text{ ft} \times 12,000 \text{ ft} = -24°F.$$

Thus the boiling point at 12,000 feet is 212 °F − 24°F = 188 °F.

Computer Chip Production.

15.

a. The initial value is 120 chips and the rate of change is 14 chips per hour.

b.

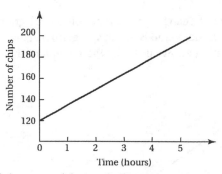

c. Recall the general form of a linear equation is

dependent variable = initial value + (rate of change × *independent variable*)

In this case, the independent variable is *time* (which we will denote t) and the dependent variable is *number of chips* produced (which we will denote N). With the initial value and rate of change given in part (a), the linear equation for this function is

Number of chips = 120 chips + (14 chips/hr × *time*),

or more compactly,

$$N = 120 + 14\,t.$$

Linear Graphs.

17.

a. The variables are (*time, diameter*).

b. The rate of change of the function is 0.2 in/yr and initial value for the function is (0 yr, 4 in), where we let $t = 0$ represent the time at which the tree was first observed. We can now draw the graph of the function:

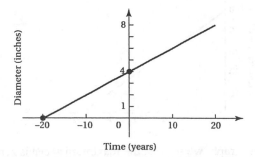

c. Extending the graph *backwards* in time, we can ask when the diameter was 0 inches. At a growth rate of 0.2 in/yr, the tree needs (4 in)/(0.2 in/yr) = 20 years to grow from 0 inches to 4 inches. So the age of the tree when it has a diameter of 4 inches is 20 years.

d. The assumption that the diameter grows linearly in time (constant growth rate) would need to be evaluated by a biologist. It may not be too accurate which will compromise the validity of this model.

19.

a. The variables in this problem are (*time, amount of sugar*).

b. The rate of change for the function is −0.1 grams/day and the initial value is (0 days, 5 grams). This information allows us to draw the graph of the function:

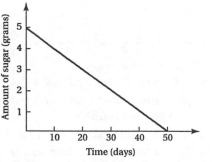

c. From the graph we see that the amount of sugar reaches zero when *time* = 50 days.

d.. A biologist would need to assess the assumption that the sugar consumption rate is constant. This may not be an accurate assumption.

21.

a. The variables in this problem are (*load, maximum speed*).

b. We are given two points on the graph of this linear function: (0 tons, 50 mph) and (20 tons, 40 mph). With these two points we can draw the linear graph of the functions:

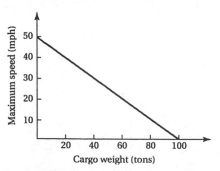

c. From the graph, we see that the maximum speed is zero when the load is 100 tons.

d. Maximum truck speeds most likely don't decrease linearly with load, but the model probably gives a rough approximation.

Linear Equations. Recall that the general form of a linear equation is

dependent variable = initial value + (rate of change × *independent variable*)

23.

a. The independent variable is *time* (or t) measured in years; we will let $t = 0$ represent January 1998. The dependent variable is *price* (or p).

b. The initial value is the price of a megabyte of memory today which is $25. The rate of change is −$5 per year (because the price decreases in time). Using the general form of the linear equation given above, the equation of this price function is

$$p = 25 - 5\,t.$$

c. July 2001 is $t = 3.5$ years after January 1998. Setting $t = 3.5$ years in the equation, we see that the price of a megabyte of memory will be

$$p = 25 - (5 \times 3.5) = \$7.50.$$

d. This function probably gives a good estimate of prices of memory over a short period of time.

25.

a. The variables in this problem are (*time, candle length*) or (t, L). We will measure time in hours with $t = 0$ representing the time at which the candle is lit.

b. The rate of change of the function is −2 cm/hr and the initial value is 20 cm. The linear equation that describes the function is

$$L = 20 - 2\,t.$$

c. At a rate of −2 cm per hour, the candle will decrease in length from 20 cm to 0 cm in (20 cm)/(2 cm/hr) = 10 hours.

d. The candle probably burns at a nearly constant rate, so this linear model is fairly accurate.

27.

a. The variables in this problem are (*miles, rental cost*) or (m, r).

b. We are told that the rate of change of rental cost with respect to miles is $0.10/mile. The initial value of the rental function is $40, the cost before any mileage is paid. The equation for the rental function is

$$r = 40 + 0.10\,m.$$

c. If you start with $90, the first $40 is paid for the flat fee. This leaves $50 for mileage. The number of miles you can drive for $50 is $50/($0.10/mile) = 500 miles.

d. This function gives a very good model of rental costs provided all of the costs are quoted correctly.

29.

a. The variables in this problem are (*time, population*) or (*t, p*) where *t* is measured in years starting with *t* = 0 in 1980.

b. The rate of change of this function is 200 people per year and one point on the graph is the initial value (0 yr, 2000 people). The population at all times is described by the equation

$$p = 2000 + 200\,t.$$

c. The year 2010 corresponds to *t* = 30 years after 1980. Setting *t* = 30 in the population equation gives a population of

$$2000 + (200 \times 30) = 2000 + 6000 = 8000 \text{ people.}$$

d. If the population grows at a constant rate, at least in an average sense, then this model is reliable.

Graphing General Linear Equations. Recall that an equation in the form $y = mx + b$ has a *y*-intercept of *b* and a slope of *m*.

31. All four graphs are shown in the figure below.

a. The equation $y = 3x - 6$ describes a straight line with *y*-intercept of (0, −6) and slope 3.

b. The equation $y = -2x + 5$ describes a straight line with *y*-intercept of (0, 5) and slope −2.

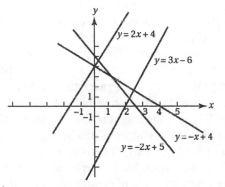

c. The equation $y = -x + 4$ describes a straight line with *y*-intercept of (0, 4) and slope −1.

d. The equation $y = 2x + 4$ describes a straight line with *y*-intercept of (0, 4) and slope 2.

Solving Linear Equations. Recall that only two steps are needed to solve a linear equation for either variable: you may add or subtract the same quantity from both sides and you may multiply both sides of the equation by the same non-zero quantity. The examples in the text illustrate the method. Units may be omitted from the calculation.

33. We are given the linear equation $P = 550 + (150 \text{ people/yr} \times t)$ where *t* = 0 represents the year 1990. To determine when the population reaches 1000 people, we must set *P* = 1000 and solve for *t* as follows.

starting equation:	$1000 \text{ people} = 550 \text{ people} + 150\frac{\text{people}}{\text{yr}} \times t$
subtract 550 people from both sides:	$(1000-550) \text{ people} = 150\frac{\text{people}}{\text{yr}} \times t$
simplify equation:	$450 \text{ people} = 150\frac{\text{people}}{\text{yr}} \times t$
divide both sides by 150 people/yr	$\dfrac{450 \text{ people}}{150\frac{\text{people}}{\text{yr}}} = t$
simplify equation:	$\dfrac{450}{150} \text{ yr} = t, \text{ or } t = 3 \text{ yr}$

We see that the population reached 1000 people in 3 years (in 1993).

35. Creating Linear Equations. In all of these problems we must find the slope of the straight line that passes through the two given points as described in the text.

a. The slope of the line through the two given points is $(7 - 13)/(2 - 4) = 3$. We now use the general form of the equation of a line

$$y = \text{initial } y + (\text{slope} \times x) = \text{initial } y + 3x$$

and substitute *either* value of the two given points to find the initial value of *y*. Using the first point, we have that

$$7 = \text{initial } y + (3 \times 2) \text{ or initial } y = 1.$$

Thus, the equation of the line passing through the two given points is $y = 3x + 1$. (Check that both points satisfy this equation.)

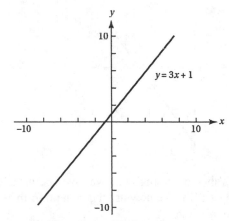

b. The slope of the line through the two given points is $(3 - 7)/(-1 - 0) = 4$. We now use the general form of the equation of a line

$$y = \text{initial } y + (\text{slope} \times x) = \text{initial } y + 4x$$

and substitute *either* value of the two given points to find the initial value of *y*. Using the first point, we have that

$3 = $ initial $y + (4 \times (-1))$ or initial $y = 7$.

Thus, the equation of the line passing through the two given points is $y = 4x + 7$. (Check that both points satisfy this equation.)

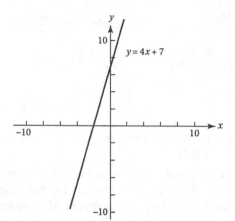

c. The slope of the line through the two given points is $(2 - (-20))/(14 - (-4)) = 22/18 = 11/9$. We now use the general form of the equation of a line

$$y = \text{initial } y + (\text{slope} \times x) = \text{initial } y + (11/9)x$$

and substitute *either* value of the two given points to find the initial value of y. Using the first point, we have that

$2 = $ initial $y + ((11/9) \times 14)$ or initial $y = -136/9$.

Thus, the equation of the line passing through the two given points is $y = (11/9)x - 136/9$. (Check that both points satisfy this equation.)

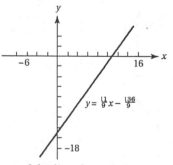

d. The slope of the line through the two given points is $(9 - 8)/(17 - 4) = 1/13$. We now use the general form of the equation of a line

$$y = \text{initial } y + (\text{slope} \times x) = \text{initial } y + (1/13)x$$

and substitute *either* value of the two given points to find the initial value of y. Using the first point, we have that

$9 = $ initial $y + ((1/13) \times 17)$ or initial $y = 100/13$.

Thus, the equation of the line passing through the two given points is $y = (1/13)x + 100/13$. (Check that both points satisfy this equation.)

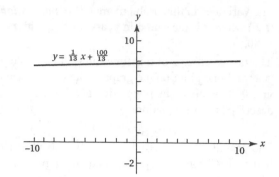

37. Linear Growth. We are looking for a function (*age, height*) or just (a, h), where time a is measured in years and $a = 0$ is the moment of birth. The variable height h is measured in inches. Remember that to find the formula for this linear function, we need to know the initial value of the height (the height at age zero), which is given to us as 20 inches, and the slope or rate of change of the function. To find the slope, we use the additional fact that when $a = 10$ years, the height is 4 feet or 48 inches. The two points (0,20) and (10,48) allow us to find that the slope of the function is $(48 \text{ in} - 20 \text{ in})/(10 \text{ yr} - 0 \text{ yr}) = 2.8$ in/yr. The equation for the function is now given by

$$h = 20 + 2.8\, a.$$

You can check that this formula gives us the two points we started with: when $a = 0$, we have $h = 20$ inches, and when $a = 10$, we have $h = 48$ inches. Furthermore, we can test other points in time. The linear model we have created says that when $a = 2$ years, the height is $h = 25.6$ inches; when $a = 6$ years, the height is $h = 36.8$ inches; when $a = 20$ years, the height is $h = 76.0$ inches (6 feet, 4 inches); and when $a = 50$ years, the height is $h = 160.0$ inches (13 feet, 4 inches!). The linear model may be accurate for small ages, but it clearly fails with increasing age, since everyone's rate of growth decreases in time.

39. Fund Raising Strategy. We can let *tickets sold* be the first variable (call it n) and *profit* be the second variable (call it P). We are told that the ticket price is \$5 which means that profit increases at a rate of \$5 per ticket; this is the rate of change of the function. Notice that the cost of the television set is a fixed cost of the event; it must be paid even if tickets are not sold. Effectively, the event starts \$350 "in the hole" and if zero tickets are sold, then the profit is −\$350. This fact gives us the initial value (0 tickets, −\$350). The equation for the profit function is

$$P = -\$350 + (5\$/\text{ticket} \times n)$$

We see that in order for the event to break even (which means that $P = \$0$), the number of tickets that must be sold is ($350)/($5/ticket) = 70 tickets.

41. Depreciation of Equipment. We will let *time* in years (denoted t) be the first variable and *depreciated value* (denoted v) be the second variable. When the washing machine is new it has a value of $v =\$1000$ which is the initial value of the function. We are told that the machine depreciates at a rate of –$50 per year (negative because the value decreases). Therefore, the equation for the depreciated value is

$$v = \$1000 - (\$50/\text{yr} \times t)$$

To find the time at which the value reaches zero, we set $v = \$0$ and solve for t. We find that the time of zero value is $1000/($50/\text{yr}) = 20$ years.

43. Pricing Strategies. We will let P be the total price of buying h hubcaps. If fewer than 10 hubcaps are purchased $(0 \le h < 10)$, then $P = \$30 \times h$; if between 10 and 20 hubcaps are purchased $(10 \le h < 20)$, the price is $P = \$25 \times h$; and if 20 or more hubcaps are purchased $(h \ge 20)$, then the price is $P = \$20 \times h$. The pricing seems reasonable and the overlaps between the brackets are not very wide: 9 hubcaps cost $270, 10 cost $250, and 11 cost $275, so it does not pay to buy 9 hubcaps. Similarly, 16 hubcaps cost $400, 17 cost $425, 18 cost $450, and 19 cost $475, while 20 cost $400, so it does not pay to buy 16, 17, 18, or 19 hubcaps.

45. Salesperson Strategies.

a. We will let s represent your monthly sales in dollars and E represent your monthly earnings. Under Plan A (monthly salary of $800 plus 10% commission), the linear function that describes earning in terms of sales is

$$E = \$800 + (0.1 \times s).$$

Notice that even with no sales $(s = 0)$ you still earn the monthly salary of $800. Added on to your $800 salary is 0.1 (10%) of your sales.

b. Under Plan B (straight commission of 20% of all sales), the linear function that describes earning in terms of sales is

$$E = 0.2 \times s.$$

Notice that there is no fixed salary and your earnings consist entirely of the 20% of sales. If there are no sales $(s = 0)$, then there are no earnings. The graph of both functions is given below.

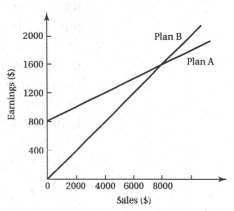

c. With sales of $2000, under Plan A , your earnings will be $E = \$800 + (0.1 \times \$2000) = \$1000$. Under Plan B your earnings will be $E = 0.2 \times \$2000 = \$400$. With sales of $2000, Plan A is preferable, because the sales are relatively low and the salary contributes most of your earnings. Under Plan A with sales of $4000, your earnings will be $E = \$800 + (0.1 \times \$4000) = \$1200$. Under Plan B with sales of $4000, your earnings will be $E = 0.2 \times \$4000 = \$800$. In this case, Plan A is still preferable. As seen from the graph, the cross-over point at which Plan B becomes preferable to Plan A is when sales are $8000; your earnings at this point are $1600. (Check that with both linear functions, your earnings are $1600.)

Unit 6C Formulas as Models

Overview

    This is an open-ended unit that is designed to show you the power of functions. It gives many examples of **nonlinear** functions — functions whose graphs are *not* straight lines. In all cases, these functions are given to you in the form of formulae involving two variables.

    The entire unit proceeds by example. You will encounter some formulae that use powers (for example, the area of a circle $A = \pi r^2$) and others that use roots (for example, the return on an investment $A = (T+1)^{Y_Y} - 1$). Two rules are given to handle functions involving powers and roots.

- we can raise both sides of an equation to the same power. For example, if it is true that $x = y$, it is also true that $x^2 = y^2$.
- we can take the same root of both sides of an equation. For example, if it is true that $x = y$, it is also true that $\sqrt{x} = \sqrt{y}$ (provided $\sqrt{x}$ and $\sqrt{y}$ are both defined).

For the ambitious ones, we take one final step and explore methods for solving equations in which a variable appears in an exponent. This requires using logarithms. For some reason, logarithms have a bad name among many people. It really isn't fair, because logarithms are useful and no more difficult to understand than many other ideas in mathematics. But like many things in mathematics and life, logarithms do require some practice to master. So if you have had bad experiences with logarithms in the past, put that behind you and start from scratch in this unit. You may also want to see Unit 12C for specific applications of logarithms in studying earthquakes, sounds, acid concentrations.

Again a few rules are needed to work with logarithms:

- taking the logarithm of a power of 10 gives the power; that is,
$$\log_{10} 10^x = x.$$
- raising 10 to a power that is the logarithm of a number gives back the number; that is,
$$10^{\log_{10} x} = x \quad (\text{for } x > 0).$$
- we can "bring down" an exponent within a logarithm by applying the *power rule* for logarithms:
$$\log_{10} a^x = x \times \log_{10} a \qquad (\text{for } a > 0)$$

Several examples illustrate the use of logarithms on practical problems. This is a challenging unit, but the ideas are powerful. It gives just another testimony to the widespread usefulness of functions.

Key Words and Phrases

nonlinear	variable	constant
participation bias	powers	roots
logarithm		

Key Concepts and Skills

- graph a nonlinear function given a table of coordinate values.
- graph a nonlinear function given a formula involving two variables and constants.
- solve an equation involving powers or roots for a specified variable.

• use logarithms to solve an equation for a variable that appears in an exponent.

Solutions 6C

3. The following graphs are plotted for values of x between -5 and 5. Other intervals can be used, although this choice seems to give a good picture of these functions.

a.

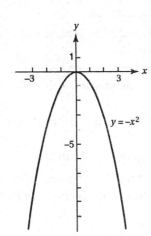

$y = -x^2$

b.

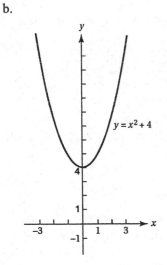

$y = x^2 + 4$

c.

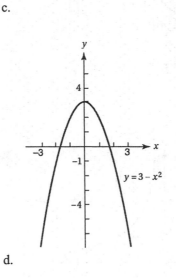

$y = 3 - x^2$

d.

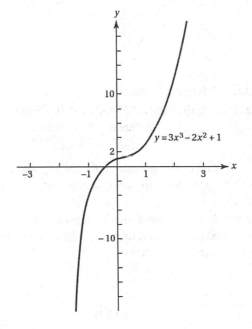

$y = 3x^3 - 2x^2 + 1$

e.

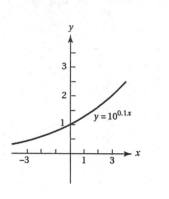

$y = 10^{0.1x}$

f.

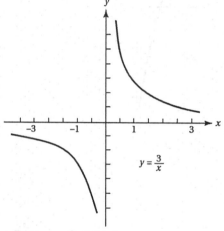

$y = \frac{3}{x}$

5. Holding Distance Constant.

a. The time required for a 400-mile trip is given by

$$\text{Time} = \frac{\text{distance}}{\text{speed}} = \frac{400 \text{ miles}}{\text{speed}}.$$

b. If we let t be the time required for the trip and v be the speed, then the relation between time and speed is $t = 400/v$.

c. The graph of the (*speed, time*) relation is shown below. Notice that as the speed increases, the time needed to travel 400 miles decreases.

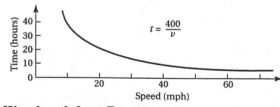

$t = \frac{400}{v}$

7. Wavelength from Frequency.

a. The fundamental relation between the speed, wavelength, and frequency of light (see Example 2) is

speed = frequency × wavelength or $c = f \times L$.

where the speed of light is always $c = 3 \times 10^8$ meter/sec. This formula can be written to give the wavelength, $L = c/f$.

b. Given that the frequency of an FM station is 106.5 megahertz = 1.065×10^8 Hertz, the wavelength is

$$L = \frac{c}{f} = \frac{3 \times 10^8 \text{ meters} / \sec}{1.065 \times 10^8 \text{ cycles} / \sec} = 2.8 \text{ meters}.$$

The wavelength of this FM radio wave is 2.8 meters — a surprisingly long wave.

c. Given that the frequency of an AM radio wave is 700 kilohertz = 7×10^5 Hertz, the wavelength is

$$L = \frac{c}{f} = \frac{3 \times 10^8 \text{ meter} / \sec}{7 \times 10^5 \text{ cycles} / \sec} = 429 \text{ meters}.$$

The wavelength of this AM radio wave is 429 meters — an even longer wave.

9. Tin Cans. The volume formula for a cylinder is useful here. A cylinder with a height of h and a radius of r has a volume of $V = \pi r^2 h$.

a. The volume of a cylinder with a radius of 2 cm = 0.02 m and a height of 12 m is

$$V = \pi (0.02 \text{ m})^2 (12 \text{ m}) = 0.015 \text{ m}^3.$$

b. Solving the volume formula for the radius (divide both sides by πh and take the square root), we see that

$$r = \sqrt{\frac{V}{\pi h}}.$$

A cylinder with a volume of 355 m³ and a height of 2 m has a radius of

$$r = \sqrt{\frac{355 \text{ m}^3}{\pi (2 \text{ m})}} = 7.5 \text{ m}.$$

Notice how the units work out!

c. Using the radius formula for a cylinder from part (b), a cylinder with a volume of 28 cm³ and a height of 12 cm has a radius of

$$r = \sqrt{\frac{28 \text{ cm}^3}{\pi (12 \text{ cm})}} = 0.86 \text{ cm}.$$

d. We can also solve the volume formula for the height. Dividing both sides of the volume formula by πr^2 gives us

$$h = \frac{V}{\pi r^2}.$$

A cylinder with a volume of 355 m³ and a radius of 25 m has a height of

$$h = \frac{355 \text{ m}^3}{\pi (25 \text{ m})^2} = 0.18 \text{ m}.$$

Notice how the units work out!

e. Using the height formula of part (d), a cylinder with a volume of 225 cm² and a radius of 8.6 mm = 0.86 cm has a height of

$$h = \frac{225 \text{ cm}^3}{\pi(0.86 \text{ cm})^2} = 96.84 \text{ cm}.$$

11. Pricing Merchandise. We must use the revenue function

$$R = -5p^2 + 40p$$

to solve the following problems.

a. If the price of the item is zero ($p = 0$), then the revenue is $R = 0$. This is consistent; if nothing is charged for the item, there can be no revenue.

b. If the price is $p = 8$, then the revenue is
$$R = -5 \times 8^2 + 40 \times 8 = \$0.$$
This says that if the price is set too high, there will be no sales and the revenue is zero.

c. With a price of $p = 3$, the revenue is
$$R = -5 \times 3^2 + 40 \times 3 = \$75.$$
With a price of $p = 5$, the revenue is also
$$R = -5 \times 5^2 + 40 \times 5 = \$75.$$

d. Some trial and error will reveal that with a price of $p = 4$, the revenue is
$$R = -5 \times 4^2 + 40 \times 4 = \$80,$$
which is the maximum revenue possible with this model.

13. Mutual Fund Return. Recall from Example 4, that if the annual return on an investment is A, then the total return over Y years is

$$T = (A+1)^Y - 1.$$

A mutual fund with an annual return of 12.2% = 0.122 has a total return over $Y = 5$ years of

$$T = (1 + 0.122)^5 - 1 = 0.778 = 77.8\%..$$

This means that an initial investment will increase in value by about 78% over 5 years.

15. Tumor Growth.

a. The initial population of tumor cells, found by setting $t = 0$, is $N = 2300$. Recall that $10^0 = 1$.

b. We must find the value of t that makes the tumor cell population equal to 1 million. The equation that must be solved for t is
$$1{,}000{,}000 = 2300 \times 10^{0.21 \times t}.$$
Dividing both sides by 2300 gives us
$$434.78 = 10^{0.21 \times t}.$$
To isolate the unknown t, we take the logarithm of both sides:
$$\log_{10} 434.78 = \log_{10}(10^{0.21 \times t}) = 0.21 \times t$$

Notice how the property $\log_{10}(10^x) = x$ has been used. Now dividing both sides by 0.21 gives us $t = 12.56$ days. In just over 12 days the tumor cell population will reach 1 million.

Solving the Compound Interest Formula.

17. To find how long it takes for an initial deposit of $P = \$2000$ to grow to $A = \$2500$ with daily compounding ($n = 365$) and an APR of 5% (0.05), we use the compound interest formula with these values and solve for Y. The equation that must be solved is

$$\$2500 = \$2000 \times (1 + \frac{0.05}{365})^{365 \times Y}.$$

Dividing both sides of the equation by \$2000, we have

$$1.25 = (1 + \frac{0.05}{365})^{365 \times Y}.$$

which must be solved for Y. Taking the logarithm of both sides, we find that $Y = 4.5$ years. It will take 4.5 years for the principal of \$2000 to increase in value to \$2500 with daily compounding and an APR of 5%.

19. If you start with an initial investment of P, it will have doubled when the accumulated balance is $2P$. To find the doubling time with daily compounding ($n = 365$) and an APR of 8%, we must solve the following equation for Y:

$$2P = P\left(1 + \frac{0.08}{365}\right)^{365 \times Y}.$$

We can cancel the common factor of P and use logarithms on both sides to solve for Y. We find that the doubling time for the account is 8.67 years.

21. A 50% growth of the principal means that $A = P + 0.5P$ or $A = 1.5P$. If we set $A = 1.5P$ in the compound interest formula with daily compounding ($n = 365$) and APR = 7%, we have the following equation:

$$1.5P = P\left(1 + \frac{0.07}{365}\right)^{365Y} = (1.0001918)^{365Y}.$$

Canceling the common factor of P and using logarithms on both sides to solve for Y, we find that $Y = 5.8$ years. It will take 5.8 years for the principal to grow by 50% of its original value with daily compounding at an APR of 7%.

23. Accelerated Loan Payment. We use the loan payment formula of Unit 5C.

a. With a principal of $P = \$25{,}000$, an APR of 9% = 0.09, and a term of $Y = 20$ years, the monthly payments on this loan will be

$$PMT = \frac{\$25{,}000 \times \left(\frac{0.09}{12}\right)}{1 - \left(1 + \frac{0.09}{12}\right)^{-12 \times 20}} = \frac{\$187.50}{0.8336} = \$224.93.$$

Paying about $225 per month, this $25,000 loan can be paid off in 20 years with an *APR* of 9%. The total amount you will pay over the term of the loan is

20 yrs × 12 payments/year × $224.93 = $53,983.

b. If you increase the monthly payments of the loan in part (a) from $224.93 to $350 per month, how long will it take to retire the loan? We again use the loan payment formula, but now treat Y as the unknown. Substituting the values above and *PMT* = $350, we have that Y satisfies the equation

$$\$350 = \frac{\$25,000 \times \left(\frac{0.09}{12}\right)}{1 - \left(1 + \frac{0.09}{12}\right)^{-12 \times Y}}.$$

Proceeding step-by-step, we can simplify this equation to

$$1.8667 = \frac{1}{1 - (1.0075)^{-12 \times Y}} \quad \text{or}$$

$$1 - (1.0075)^{-12 \times Y} = 0.5357.$$

Some more rearranging leads to

$$0.4643 = (1.0075)^{-12Y},$$

which can be solved for Y using logarithms. We discover that the term of the new loan with the accelerated payments is $Y = 8.56$ years. Increasing the monthly payments by $125, decreases the term of the loan by almost 12 years.

The total amount you will pay over the term of the loan is approximately

8.6 yrs × 12 payments/year × $350 = $36,120,

a significant savings!

25. Biweekly Mortgage Payments.

a. With a 30-year mortgage and an APR of 9%, the monthly payments on a loan of $100,000 are

$$PMT = \frac{\$100,000 \times \left(\frac{0.09}{12}\right)}{1 - \left(1 + \frac{0.09}{12}\right)^{-12 \times 30}} = \$804.62.$$

Over the course of a year you pay

12 payments/year × $804.62 = $9655.

b. Half of your regular monthly payments is $402.31 which is paid 26 times per year. Thus, instead of paying 12 × $804.62 = $9655 per year (under the original payment plan), you pay 26 × $402.31 = $10,460. This additional $805 accelerates the loan payment plan.

c. We need to start with the loan payment formula and apply it to a $100,000 loan with an *APR* of 9%, $n = 26$ payments per year, and payments of $402. Then we solve for the number of years Y that the loan will last. Substituting these values into the loan formula, we have

$$\$402 = \frac{\$100,000 \times \left(\frac{0.09}{26}\right)}{1 - \left(1 + \frac{0.09}{26}\right)^{-26 \times Y}},$$

A few steps of simplification give us

$$1 - \left(1 + \frac{0.09}{26}\right)^{-26 \times Y} = \frac{\$100,000 \times \left(\frac{0.09}{26}\right)}{\$402} = 0.860,$$

Because the unknown Y appears in the exponent, logarithms must be applied to both sides of the equation. The eventual result is that the term of the loan is reduced from 30 years to $Y = 21.9$ years.

d. There is a clear advantage in paying off the loan early. With monthly payments for 30 years, you pay a total of 30 years × 12 payments/year × $804/payment = $289,440. With biweekly payments for 22 years, you pay a total of 22 years × 26 payments/year × $402/payment = $229,944. There is the disadvantage that an additional $800 per year must be paid, and that payments must be made twice a month.

27. Probability. We are given that the probability of rolling n 1's in a row on a die is $p = 6^{-n}$. The chances of rolling three 1's in a row is $6^{-3} = 1/6^3 = 1/216 = 0.0046$. The chances of rolling seven 5's in a row is the same as rolling seven 1's in a row which is $6^{-7} = 1/6^7 = 0.0000036$.

P 29. Postage Inflation. The formula

$$p = 3 \times 2^{t/10}$$

predicts that the postage in 1997 ($t = 42$) is $3 \times 2^{42/10} = 55$ cents. The formula is not very accurate.

7. EXPONENTIAL GROWTH AND DECAY

Overview

If you had to learn just one lesson from a quantitative reasoning course, it might well be the difference between linear growth (as studied in Unit 6B) and exponential growth which is the subject of this chapter. Exponential growth and decay impacts our everyday lives in so many ways, but most people are not aware of its presence or its power. In this chapter you will learn how everything from bank accounts to populations to radioactive waste are governed by exponential models; and you will learn how to construct exponential models to describe many everyday phenomena. In practical terms, this is an immensely important chapter.

Unit 7A Exponential Astonishment

Overview

The difference between linear and exponential growth is stated on the first page of the chapter.

- *Linear growth* occurs when a quantity grows by the same *absolute* amount in each unit of time.
- *Exponential growth* occurs when a quantity grows by the same *relative* amount — that is, by the same *percentage* — in each unit of time.

Notice how these facts are related to the ideas of absolute and relative change studied in Unit 4B. It would be a good idea to contemplate these two statements and try to understand what they really mean. Hopefully this unit will also help!

This unit introduces the ideas surrounding exponential growth and decay in a very basic and accessible way. We use three parables (*From Hero to Headless*, *The Magic Penny*, and *Bacteria in a Bottle*) that illustrate quite dramatically the power of exponential growth. The goal is to develop some intuition about exponential growth and understand doubling processes.

The lessons of the unit are summarized in the highlight box at the end of the unit:

- Exponential growth is characterized by repeated doublings. With each doubling the amount of increase is approximately equal to the *sum* of all preceding doublings.

- Exponential growth cannot continue indefinitely. After only a relatively small number of doublings, exponentially growing quantities reach impossible proportions.

Key Words and Phrases

linear growth   exponential growth   doublings

Key Concepts and Skills

- explain the difference between linear and exponential growth.
- identify whether a given growth pattern is linear or exponential.
- understand the implications of a doubling process.

Solutions Unit 7A

1. Growth and Decay in the News. Solutions will vary.

3. Linear Versus Exponential.

a. Because the *absolute* growth rate (230 people per year) is constant, the population of Winesburg is growing linearly.

b. Because the *relative* or *percentage* growth rate (−12% per year) is constant, the price of computer memory is decreasing exponentially.

c. Because the *relative* or *percentage* rate (−1.5% per year) is constant, the number of births is declining exponentially.

d. Because the *absolute* rate (−$200 per year) is constant, the value of the equipment is declining linearly.

5. Chessboard Wheat. According to the story and Table 7-1, square #20 of the chess board has $2^{19} = 524,288$ grains of wheat. The total number of grains on the board at this point would be $2^{20} - 1 = 1,048,575$; over one million grains of wheat!

7. The Weight of All That Grain.

a. The weight of the grain on the chessboard is

$$\left(1.8 \times 10^{19} \text{ grains}\right) \times \left(\frac{1 \text{ pound}}{7000 \text{ grains}}\right) \times \left(\frac{1 \text{ ton}}{2000 \text{ pounds}}\right)$$

$$= 1.3 \times 10^{12} \text{ tons.}$$

b. The ratio of the grain on the chessboard to the annual world grain harvest is $(1.3 \times 10^{12} \text{ tons})/(2 \times 10^9 \text{ tons}) = 650$. The grain on the chessboard is 650 times or 65,000% of the annual world grain harvest

9. Bacteria in a Bottle.

a. Table 7-3 shows that one minute before 12:00 the bottle is 1/2 full; two minutes before 12:00 the bottle is $1/4 = 1/(2^2)$ full; three minutes before 12:00 the bottle is $1/8 = 1/(2^3)$ full; and so on. Therefore, 5 minutes before 12:00 at 11:55, the bottle is $1/(2^5)$ full or 1/32 full. The number of bacteria at 5 minutes before 12:00 (which is 55 minutes after 11:00) is 2^{55}.

b. Table 7-3 shows that 55 minutes before 12:00 at 11:05, the bottle is $1/2^{55}$ full or 2.8×10^{-17} full. The number of bacteria at 55 minutes before 12:00 (which is 5 minutes after 11:00) is $2^5 = 32$.

11. A Layer of Bacteria. Because the total surface area of the Earth is about 5×10^{14} square meters, we can calculate the thickness of the layer of bacteria on the Earth simply by dividing:

$$\frac{1.3 \times 10^{15} \text{ m}^3}{5.1 \times 10^{14} \text{ m}^2} = 2.5 \text{ m.}$$

Notice how the units work out! The bacteria layer would be 2.5 meters thick over the entire Earth

13. Population Doubling.

a.

Year	Population	Year	Population
2000	6×10^9	2550	1.229×10^{13}
2050	1.2×10^{10}	2600	2.458×10^{13}
2100	2.4×10^{10}	2650	4.915×10^{13}
2150	4.8×10^{10}	2700	9.830×10^{13}
2200	9.6×10^{10}	2750	1.966×10^{14}
2250	1.92×10^{11}	2800	3.932×10^{14}
2300	3.84×10^{11}	2850	7.864×10^{14}
2350	7.68×10^{11}	2900	1.573×10^{15}
2400	1.536×10^{12}	2950	3.146×10^{15}
2450	3.072×10^{12}	3000	6.291×10^{15}
2500	6.144×10^{12}		

b. We see that there will be 5.1×10^{14} people on the Earth occupying 5.1×10^{14} m^2 sometime between 2800 and 2850.

c. The number of people that could be supported assuming each person needs 10^4 m^2 is

$$\frac{5.1 \times 10^{14} \text{ m}^2}{10^4 \text{ m}^2 / \text{person}} = 5.1 \times 10^{10} \text{ people.}$$

This limit would be reached shortly after the year 2150.

Unit 7B Doubling Time and Half-Life

Overview

Exponential growth is characterized by a constant **doubling time**. If a quantity (for example, a population or a bank account) is growing exponentially, then it doubles its size during a fixed period of time, and it continues to double its size over that same period of time forever. For example, if a tumor, growing exponentially, has a doubling time of two months, then it doubles its size during the first two months, and doubles its size again during the next two months, and continues to double in size *every* two months. Knowing the doubling time essentially defines the growth pattern for all times.

We denote the doubling time T_{double}. If we know T_{double} for a particular quantity that grows exponentially, then over a period of t time units, the quantity will increase by a factor of

$$2^{t/T_{\text{double}}}.$$

If we know the doubling time and the initial value of a particular quantity that grows exponentially, then we can find its value at all later times. The new values at later times are given by

$$\text{new value} = \text{initial value} \times 2^{t/T_{\text{double}}}.$$

If we know that a quantity grows with a constant percentage growth rate (for example, 5% per year), then we know it grows exponentially and that it has a constant doubling time. This suggests that there should be a connection between the percentage growth rate, which we call P, and the doubling time. We use a specific example of an exponentially growing population to present a widely used formula that relates percentage growth rate to doubling time. It is called the **Rule of 70** or the **Approximate Doubling Time Formula**. It says that

$$T_{double} \approx \frac{70}{P}.$$

This formula is an approximation and works best when the percentage growth rate is small (say, less that 10%). For example, if a bank account grows at 4% per year, it will double it value in approximately 70/4 = 17.5 years.

Everything we learned about exponential growth has a parallel with exponential decay. For example, if a quantity decays exponentially at a rate of 5% per month, it *decreases* by 5% during the first month, and by 5% during the second month, and continues to decrease by 5% every month. A quantity that decays exponentially has a constant **half-life** – the period of time over which it decreases its size by 50% or one-half.

We denote the doubling time T_{half}. If we know T_{half} for a particular quantity that decays exponentially, then over a period of t time units, the quantity will increase by a factor of

$$\left(\frac{1}{2}\right)^{t/T_{half}}$$

If we know the half-life and the initial value of a particular quantity that grows exponentially, then we can find its value at all later times. The new values at later times are given by

$$\text{new value} = \text{initial value} \times \left(\frac{1}{2}\right)^{t/T_{half}}.$$

The **Rule of 70** or **Approximate Half-Life Formula** also applies to exponentially decaying quantities. If a quantity decreases by P% per unit time, the then half-life is given by

$$T_{1/2} \approx \frac{70}{P}.$$

As with the Approximate Doubling Time Formula, this half-life formula is approximate and works best when P is small (say, less than 10%).

You might wonder if there are exact formulae for finding the doubling time or half-life from the percentage growth or decay rates. For those who are curious, the unit closes with the *exact* doubling time and half-life formulae. These formulae are a bit more complicated and require the use of logarithms. They are exact for all percentage growth and decay rates, not just small ones.

Key Words and Phrases

doubling time	percentage growth rate	approximate doubling time formula
Rule of 70	half-life	percentage decay rate
approximate half-life formula	exact doubling time formula	exact half-life formula

Key Concepts and Skills

- identify the percentage growth rate from the description of an exponential growth process.
- find the doubling time from percentage growth rate.
- find the percentage growth rate from doubling time.
- determine the value of exponentially growing quantity given the doubling time and initial value.
- identify the percentage decay rate from the description of an exponential decay process.
- find the half-life from percentage decay rate.
- find the percentage decay rate from half-life.
- determine the value of exponentially decay quantity given the half-life and initial value.

Solutions Unit 7B

1. Change and Doubling Time.

a. There are eight three-hour intervals in 24 hours, so eight doublings will occur in 24 hours. This means that the population increases by a factor of $2^8 = 256$ in 24 hours.

There are 7 days/week $\times$ 8 doublings/day = 56 three-hour intervals per week, so 56 doublings will occur in one week. This means that the population increases by a factor of $2^{56} = 7.2 \times 10^{16}$ in a week.

b. There are three 10-year intervals in 30 years, so three doublings will occur in 30 years. This means that the population increases by a factor of $2^3 = 8$ in 30 years.

There are five 10-year intervals in 50 years, so five doublings will occur in 50 years. This means that the population increases by a factor of $2^5 = 32$ in 50 years.

c. A quadrupling is two doublings, so it will take two doubling times or 44 years for the population to quadruple.

3. Growth from Doubling Time.
In these problems we must use the growth formula

$$\text{new value} = \text{initial value} \times 2^{t / T_{\text{double}}}$$

a. We set initial value = \$500 and T_{double} = 15 years. After t = 20 years, the balance will be

$$\text{new value} = \$500 \times 2^{20/15} = \$500 \times 2^{1.3333}$$
$$= \$500 \times 2.52 = \$1260.$$

After 30 years, the money has been in the account for two doubling times, so the balance will be $4 \times \$500 = \$2000$. Check that the growth formula gives the same result!

b. We set initial value = 15,600 and T_{double} = 8 years. After t = 12 years, the population will be

$$\text{new value} = 15,600 \times 2^{12/8} = 15,600 \times 2^{1.5}$$
$$= 15,600 \times 2.83 = 44,123.$$

After 24 years, the population has been growing for three doubling times, so the population will be $8 \times 15{,}600 = 124{,}800$. Check that the growth formula gives the same result!

c. We set initial value = 1 and $T_{double} = 1.5$ months. After $t = 3$ years = 36 months, cell population will be

$$\text{new value} = 1 \times 2^{36/1.5} = 1 \times 2^{24} = 16{,}777{,}216.$$

After $t = 4$ years = 48 months, the cell population will be

$$\text{new value} = 1 \times 2^{48/1.5} = 1 \times 2^{32} = 4{,}294{,}967{,}296$$

or about 4 billion cells.

5. World Population Growth. Let's take the initial value to be the 1990 population of 5.2 billion. Setting $T_{double} = 40$ years, the growth formula is

$$\text{new value} = 1 \times 2^{t/40}.$$

After $t = 20$ years (for the year 2010 which is half of a doubling time in the future), the population will be

$$\text{new value} = 5.2 \text{ billion} \times 2^{20/40} = 5.2 \text{ billion} \times 1.4142$$
$$= 7.35 \text{ billion}.$$

After $t = 70$ years (for the year 2060 which is about one and a half of a doubling times in the future), the population will be

$$\text{new value} = 5.2 \text{ billion} \times 2^{70/40} = 5.2 \text{ billion} \times 3.3636$$
$$= 17.49 \text{ billion}.$$

After $t = 110$ years (for the year 2100 which is almost three doubling times in the future), the population will be

$$\text{new value} = 5.2 \text{ billion} \times 2^{110/40} = 5.2 \text{ billion} \times 6.7271$$
$$= 34.98 \text{ billion}.$$

7. Doubling Time Practice. In all of these problems, the approximate relation between the growth rate and the doubling time can be used. Recall from the text that if r is the fractional growth rate and $P = 100 \times r$ is the corresponding percentage growth rate, then the doubling time is given by $T_{double} \approx 70/P$. The approximation is very good when P is small, about 10% or less. If the doubling time T_{double} is known then in a period of time t, a quantity will increase by a factor of $2^{t/T_{double}}$.

a. With a percentage growth rate of 0.7% per month, the doubling time is approximately $70/P = 70/0.7 = 100$ months. After one year (12 months), the prices will increase by a factor of $2^{12/100} = 1.09$. After eight years (96 months), the prices will increase by a factor of $2^{96/100} = 1.95$.

b. With a percentage growth rate of 1.6% per year, the doubling time is approximately $70/P = 70/1.6 = 44$ years. After 10 years, the population will increase by a factor of $2^{10/44} = 1.17$, making the population 6 billion $\times$ 1.17 = 7 billion. After 100 years, the population will increase by a factor of $2^{100/44} = 4.83$, making the

population 6 billion $\times$ 4.83 = 29 billion. After 1000 years, the population will increase by a factor of $2^{1000/44}$ which is almost 7 million, making the population 6 billion $\times$ 7 million = 4.2×10^{16}. Clearly, such growth is impossible.

c. With a percentage growth rate of 1.9% per year, the doubling time is approximately $70/P = 70/1.9 = 37$ years. After one decade (10 years), oil consumption will increase by a factor of $2^{10/37} = 1.21$.

9. Change and Half-Life.

a. There are two 35-year periods in 70 years, so two halvings will occur in 70 years. This means that the amount of material decreases to $(1/2)^2 = 1/4$ of its size in 70 years. There are four 35-year periods in 140 years, so four halvings will occur in 140 years. This means that the amount of material decreases to $(1/2)^4 = 1/16$ of its size in 140 years.

b. There are four 10-year periods in 40 years. This means that the population decreases to $(1/2)^4 = 1/16$ of its size in 40 years. There are seven 10-year periods in 70 years. This means that the population decreases to $(1/2)^7 = 1/128$ of its size in 70 years.

c. There are two 12-hour periods in 24 hours. This means that the amount of drug decreases to $(1/2)^2 = 1/4$ of its size in 24 hours. There are three 12-hour periods in 36 hours. This means that the amount of drug decreases to $(1/2)^3 = 1/8$ of its size in 36 hours.

11. Decay from Half-Life. In these problems we must use the decay formula

$$\text{new value} = \text{initial value} \times \left(\frac{1}{2}\right)^{t/T_{half}}$$

a. We set initial value = 200 grams and $T_{half} = 1000$ years. After 5550 years, the amount of the sample remaining will be

$$200 \times \left(\frac{1}{2}\right)^{5550/1000} = 200 \times \left(\frac{1}{2}\right)^{5.55} = 4.27 \text{ gm.}$$

After 10,200 years, the amount will be

$$200 \times \left(\frac{1}{2}\right)^{10{,}200/1000} = 200 \times \left(\frac{1}{2}\right)^{10.2} = 0.17 \text{ gm.}$$

b. We set initial value = 2500 trees and $T_{half} = 15$ years. After 35 years, the number of trees remaining will be

$$2500 \times \left(\frac{1}{2}\right)^{35/15} = 2500 \times \left(\frac{1}{2}\right)^{2.33} = 496 \text{ trees.}$$

After 90 years, the number of trees remaining will be

$$2500 \times \left(\frac{1}{2}\right)^{90/15} = 2500 \times \left(\frac{1}{2}\right)^{6} = 39 \text{ trees.}$$

c. We set initial value = 100 mg and $T_{half} = 20$ hours. After 30 hours, the amount of drug remaining will be

$$100 \times \left(\frac{1}{2}\right)^{30/20} = 100 \times \left(\frac{1}{2}\right)^{1.5} = 35.35 \text{ mg.}$$

After 70 hours, the amount of drug remaining will be

$$100 \times \left(\frac{1}{2}\right)^{70/20} = 100 \times \left(\frac{1}{2}\right)^{3.5} = 8.84 \text{ mg.}$$

13. Half-Life Practice. In these problems, the approximate relation between the decay rate and half-life can be used. Recall that if r is the fractional decay rate and $P = 100 \times r$ is the corresponding percentage decay rate, then $T_{half} \approx 70/P$. The approximation is very good when the percentage decay rate is small, about 10% or less.

a. We are given that the percentage decay rate is 2% per year. Therefore, the half-life is approximately $70/P = 70/2 = 35$ years. In $t = 5$ years the CPI will decrease by a factor of $(1/2)^{5/35} = (1/2)^{0.14} = 0.91$, or to 0.91 of its original value.

b. We are given that the percentage decay rate is 3.3% per year. Therefore, the half-life is approximately $70/P = 70/3.3 = 21.2$ years. In $t = 5$ years the population will decrease by a factor of $(1/2)^{5/21.2} = (1/2)^{0.24} = 0.85$, or to 0.85 of its original value. In $t = 50$ years the population will decrease by a factor of $(1/2)^{50/21.2} = (1/2)^{2.36} = 0.19$, or to 0.19 of its original value.

c. We are given that the percentage decay rate is 10% per hour. Therefore, the half-life is approximately $70/P = 70/10 = 7$ hours. In $t = 10$ hours the amount of drug will decrease by a factor of $(1/2)^{10/7} = (1/2)^{1.43} = 0.37$, or to 0.37 of its original value. In $t = 24$ hours the amount of drug will decrease by a factor of $(1/2)^{24/7} = (1/2)^{3.43} = 0.09$, or to 0.09 of its original value.

d. We are given that the percentage decay rate is 6% per year. Therefore, the half-life is approximately $70/P = 70/6 = 11.7$ years. This means that the population decreases to 50% or one-half of its present population in 11.7 years.

e. With a decay rate of 0.0005% per year, the half-life is approximately $70/P = 70/0.0005 = 140,000$ years. After 100,000 years the substance will decrease in mass by a factor of $(1/2)^{100,000/140,000} = 0.61$ giving it a mass of 100 kg $\times 0.61 = 61$ kg.

15. Using the Formulas.

a. If a quantity doubles every four days, its doubling time is four days! The result is exact.

b. The quantity must double in the first two days and double again in the next two days (giving a four-fold increase in four days). Therefore, its doubling time is two days. The result is exact.

c. The half-life of a quantity that halves it size every four days is four days! The result is exact.

d. With a percentage growth rate of 5% per year, the doubling time is approximately $70/P = 70/5 = 14$ years. The approximation is good.

e. With a percentage decay rate of 5% per year, the half-life is approximately $70/P = 70/5 = 14$ years. The approximation is good.

f. With a percentage decay rate of 5% per hour, the half-life is approximately $70/P = 70/5 = 14$ hours. The approximation is good.

17. Working with Growth Rates.

a. This statement is true because 10% of $1000 is $100, meaning prices rise from $1000 to $1100.

b. Three percent of 100 million is three million so the statement is true.

c. This statement is true. Two months ago prices were $100. After a 100% increase during the next month, prices doubled to $200. After another 100% increase during the next month, prices again doubled to $400.

d. This statement is not precisely true. Notice that if a quantity increases by 1% per month for 12 months the total increase is by a factor of $(1.01)^{12} = 1.1268 = 1 + 0.1268 = 100\% + 12.68\%$. We see that the increase over a year is actually 12.68%. This is the effect of compounding. So a $100 deposit would not increase to exactly $112 in a year.

P 18. Exact Formula for the Doubling Time. In these problems, the exact relation between the growth rate and the doubling time should be used. Recall from the text that if r is the fractional growth rate and $P = 100 \times r$ is the corresponding percentage growth rate, then $T_{double} = \log 2/\log(1 + r)$. We will use log base 10 and the fact that $\log_{10} 2 = 0.3$.

7a. A percentage growth rate of 0.7% per month corresponds to a fractional growth rate of $r = 0.007$ per month. Therefore, the exact doubling time is

$$T_{double} = \frac{\log 2}{\log(1+r)} = \frac{0.3}{\log(1+0.007)} = 99.03 \text{ months.}$$

After one year (12 months), the prices will increase by a factor of $2^{12/99.03} = 1.087$. After eight years (96 months), the prices will increase by a factor of $2^{96/99.03} = 1.958$. The agreement with the approximate formula is very good.

7b. A percentage growth rate of 1.6% per year corresponds to a fractional growth rate of $r = 0.016$ per year. Therefore, the exact doubling time is

$$T_{double} = \frac{\log 2}{\log(1+r)} = \frac{0.3}{\log(1+0.016)} = 43.52 \text{ years.}$$

The agreement with the approximate formula is very good.

7c. A percentage growth rate of 1.9% per year corresponds to a fractional growth rate of $r = 0.019$ per year. Therefore, the exact doubling time is

$$T_{\text{double}} = \frac{\log 2}{\log(1+r)} = \frac{0.3}{\log(1+0.019)} = 36.70 \text{ years.}$$

The agreement with the approximate formula is very good.

P 19. Exact Formula for the Half-Life. In these problems, the exact relation between the decay rate and half-life should be used. Recall that if r is the fractional decay rate and $P = 100 \times r$ is the corresponding percentage decay rate, then

$$T_{\text{half}} = -\log 2/\log(1-r).$$

We will use log base 10 and the fact that $\log_{10} 2 = 0.3$.

13a. A percentage decay rate of 2% per year corresponds to a fractional decay rate of $r = 0.02$ per year. Therefore, the exact half-life is

$$T_{\text{half}} = -\frac{\log 2}{\log(1-r)} = -\frac{0.3}{\log(1-0.02)} = 34.19 \text{ years.}$$

In $t = 5$ years the CPI will decrease by a factor of $(1/2)^{5/34.19} = (1/2)^{0.20} = 0.904$, or to 0.904 of its original value. The agreement with the approximate formula is very good.

13b. A percentage decay rate of 3.3% per year corresponds to a fractional growth rate of $r = 0.033$ per year. Therefore, the exact half-life is

$$T_{\text{half}} = -\frac{\log 2}{\log(1-r)} = -\frac{0.3}{\log(1-0.033)} = 20.59 \text{ years.}$$

In $t = 5$ years the population will decrease by a factor of $(1/2)^{5/20.59} = (1/2)^{0.24} = 0.845$. In $t = 50$ years the population will decrease by a factor of $(1/2)^{50/20.59} = (1/2)^{2.36} = 0.186$. The agreement with the approximate formula is very good.

13c. A percentage decay rate of 10% per hour corresponds to a fractional growth rate of $r = 0.1$ per hour. Therefore, the exact half-life is

$$T_{\text{half}} = -\frac{\log 2}{\log(1-r)} = -\frac{0.3}{\log(1-0.1)} = 6.56 \text{ hours.}$$

In $t = 10$ hours the amount of drug will decrease by a factor of $(1/2)^{10/6.56} = (1/2)^{1.52} = 0.35$. In $t = 24$ hours the amount of drug will decrease by a factor of $(1/2)^{24/6.56} = (1/2)^{3.66} = 0.079$. The agreement with the approximate formula is very good.

13d. A percentage decay rate of 6% per year corresponds to a fractional growth rate of $r = 0.06$ per year. Therefore, the exact half-life is

$$T_{\text{half}} = -\frac{\log 2}{\log(1-r)} = -\frac{0.3}{\log(1-0.06)} = 11.16 \text{ hours.}$$

This means that the population decreases to 50% or one-half of its present population in 11.16 years. The agreement with the approximate formula is very good.

13e. A percentage decay rate of 0.0005% per year corresponds to a fractional growth rate of $r = 0.000005$ per year. Therefore, the exact half-life is

$$T_{\text{half}} = -\frac{\log 2}{\log(1-r)} = -\frac{0.3}{\log(1-0.000005)} = 138,155 \text{ years.}$$

After 100,000 years the substance will decrease in mass by a factor of $(1/2)^{100,000/138,155} = 0.605$ giving it a mass of 100 kg $\times$ 0.605 = 60.5 kg. The agreement with the approximate formula is very good.

21. Investment Doubling Times.

a. The *APY* of 8.50% is the percentage growth rate P of the account. The approximate doubling time is $70/P = 70/8.50 = 8.24$ years. Noting that the fractional growth rate is $r = 0.085$, the exact doubling time is $\log 2/\log(1.085) = 8.50$ years. The agreement between the two formulas is good.

b. The *APY* of 4.65% is the percentage growth rate P of the account. The approximate doubling time is $70/P = 70/4.65 = 15.05$ years. Noting that the fractional growth rate is $r = 0.0465$, the exact doubling time is $\log 2/\log(1.0465) = 15.25$ years. The agreement between the two formulas is good.

Unit 7C Exponential Modeling

Overview

Having learned about the fundamentals of exponential growth and decay, we can put these ideas to work. Just as we used linear functions to model real world situations in Chapter 6, we will now use exponential functions to model situations in which exponential growth or decay occur. We begin by introducing a general **exponential growth law**

$$Q = Q_0 \times (1 + r)^t,$$

and a parallel **exponential decay law**

$$Q = Q_0 \times (1 - r)^t.$$

Both of these laws describe how the quantity Q grows or decays with respect to time, which is denoted t. These laws require an initial value Q_0 and a growth or decay rate r. Notice that r is always positive and that the units used for r and t must be the same (for example, if t has units of *days*, then r has units of $1/days$). The only difference between the two laws is a plus or minus sign in front of r. Once the growth or decay law is found, it can be used to predict the value of Q at all future times.

Be sure to study the highlight box entitled Summary of Exponential Growth and Decay Laws. It shows that there are really two forms for the growth law depending on whether you are given a growth rate or a doubling time. Similarly, there are two forms for the decay law depending on whether you are given a decay rate or a half-life.

With these two laws at hand, the rest of the unit is devoted to various applications. We look at how population growth, economic inflation, oil consumption, pollution, and drugs in the blood can all be modeled using these laws. Of particular importance is the technique called **radioactive dating**, which also relies on the exponential decay law. If these examples don't convince you of the widespread presence of exponential growth and decay, you will find even more applications in the problems!

Key Words and Phrases

exponential growth law   exponential decay   radioactive dating
                          law

Key Concepts and Skills

- given either a growth rate or a doubling time, use the appropriate form of the exponential growth law to model an exponentially growing quantity.
- given either a decay rate or a half-life, use the appropriate form of the exponential decay law to model an exponentially decaying quantity.
- understand economic inflation and know how to determine inflation-adjusted costs.
- understand radioactive dating and know how to determine the age of a material that contains a radioactive element.

Solutions Unit 7C

1. Linear Versus Exponential Population Growth.

a, b. The growth of Linear City, with a population of 100,000 in 1990 and growing at a constant rate of 10,000 people per year is shown in the second column of the table below. The growth of Exponential City, with a population of 100,000 in 1990 and growing at a constant percentage rate of 10% per year is shown in the third column of the table below.

Year	Linear Population	Exponential Population
1990	100,000	100,000
1991	110,000	110,000
1992	120,000	121,000
1993	130,000	133,100
1994	140,000	146,410
1995	150,000	161,051
1996	160,000	177,156
1997	170,000	194,872
1998	180,000	214,359
1999	190,000	235,795
2000	200,000	259,374
2001	210,000	285,312
2002	220,000	313,843
2003	230,000	345,227
2004	240,000	379,750
2005	250,000	417,725

The graphs of both populations are shown below.

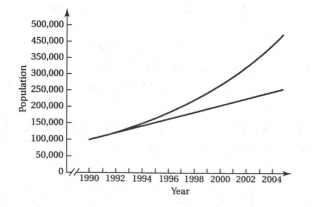

c. Linear City grows with a constant *absolute* growth rate which is the property of linear growth. The graph of its population is a straight line. Exponential City grows with a constant *percentage* growth rate which is the property of exponential growth. The graph of its

population is a steeply rising curve. Growth rates are rarely constant for 15 years, but both linear and exponential growth models can give good approximations to real population growth.

Exponential Growth and Decay Laws. In these problems, the general exponential growth (or decay) law should be used. If a quantity Q has an initial value of Q_0 and *grows* exponentially at a fractional rate of r per year, then after t years the value of Q is

$$Q = Q_0 \times (1 + r)^t.$$

If a quantity Q has an initial value of Q_0 and *decays* exponentially at a fractional rate of r per year, then after t years the value of Q is

$$Q = Q_0 \times (1 - r)^t.$$

3.

a. Letting Q represent the population of the town, we are given that the initial population is $Q_0 = 85,000$. The growth rate is $r = 0.024$ per year (corresponding to a percentage growth rate of 2.4% per year). The growth law for the population is

$$Q = 85,000 \times (1 + 0.024)^t = 85,000 \times (1.024)^t.$$

b.

Year	Population
0	85,000
1	87,040
2	89,129
3	91,268
4	93,458
5	95,701
6	97,998
7	100,350
8	102,759
9	105,225
10	107,750

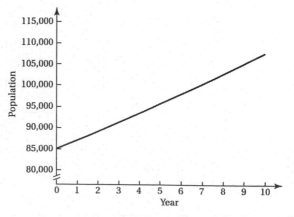

5.

a. Letting Q stand for the number of crimes, we are given that the initial value is $Q_0 = 800$ (in 1995). The growth rate is $r = 0.03$ per year (corresponding to a percentage growth rate of 3% per year). The growth law for the crimes is

$$Q = 800 \times (1 + 0.03)^t = 800 \times (1.03)^t.$$

b.

Year	Homicides
0	800
1	824
2	849
3	874
4	900
5	927
6	955
7	984
8	1013
9	1043
10	1075

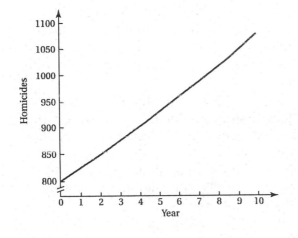

7.

a. Letting Q stand for the town's population, we are given that the initial value is $Q_0 = 10,000$ people. The decay rate is $r = 0.003$ per year (corresponding to 0.3% per year). The decay law for the population is

$$Q = 10,000 \times (1 - 0.003)^t = 10,000 \times (0.997)^t.$$

b.

Month	Population
0	10,000
1	9970
2	9940
3	9910
4	9881
5	9851
6	9821
7	9792
8	9763
9	9733
10	9704

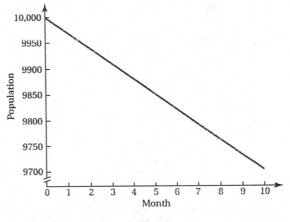

9.

a. Letting Q stand for the value of the peso, we are given that the initial value is $Q_0 = \$0.25$. The decay rate is $r = 0.1$ per week (corresponding to 10% per week). The decay law for the pesos is

$$Q = \$0.25 \times (1 - 0.1)^t = \$0.25 \times (0.9)^t.$$

b. The values in the table are rounded to the nearest cent.

Week	Value
0	$0.25
1	$0.23
2	$0.20
3	$0.18
4	$0.16
5	$0.15
6	$0.13
7	$0.12
8	$0.11
9	$0.10
10	$0.09

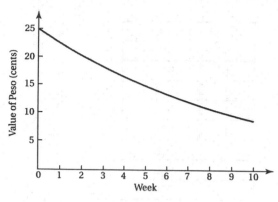

11.

a. Letting Q stand for your monthly salary, we are given that the initial value is $Q_0 = \$2000$. The growth rate is $r = 0.05$ per year (corresponding to 5% per year). The growth law for your salary is

$$Q = \$2000 \times (1 + 0.05)^t = \$2000 \times (1.05)^t .$$

b.

Year	Monthly Salary
0	$2000
1	$2100
2	$2205
3	$2315
4	$2431
5	$2553
6	$2680
7	$2814
8	$2955
9	$3103
10	$3258

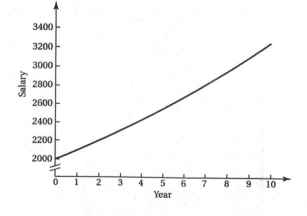

13. Metropolitan Population Growth. We can let $t = 0$ be 1990 and $Q_0 = 85,000$ be the 1990 population. The

annual percentage growth rate of 3% correspond to $r = 0.03$ per year. The growth law is given by

$$Q = 85,000 \times (1 + 0.03)^t = 85,000 \times (1.03)^t.$$

In 2010 ($t = 20$) the population according to the model will be

$$Q = 85,000 \times (1.03)^{20} = 153,519.$$

In 2100 ($t = 110$) the population according to the model will be

$$Q = 85,000 \times (1.03)^{110} = 2,195,400.$$

15. Effects of Inflation. For all three problems, the annual percentage growth rate is 3%, so $r = 0.03$ per year, and growth law is

$$Q = Q_0 \times (1.03)^t,$$

where $t = 0$ corresponds to 1990 and $t = 8$ corresponds to 1998.

a. The 1998 price of groceries that had a price of $100 in 1990 is

$$Q = \$100 \times (1.03)^8 = \$126.68.$$

b. The 1998 tuition that was $4100 in 1990 is

$$Q = \$1400 \times (1.03)^8 = \$1773.48.$$

c. The cost of the 1998 concert ticket that cost $35 in 1990 is

$$Q = \$35 \times (1.03)^8 = \$44.34.$$

17. Annual Inflation Rates from Monthly Rates.

a. At a monthly growth rate of 0.6% per month ($r = 0.006$ per month), in 12 months any quantity will increase by a factor of

$$(1 + 0.006)^{12} = 1.006^{12} = 1.074.$$

Thus the annual growth rate is $r = 0.074$ or 7.4% which is slightly greater than 12 times the monthly growth rate.

b. In 12 months, at a monthly decay rate of 0.8% per month ($r = 0.008$ per month), any quantity will decrease by a factor of

$$(1 - 0.008)^{12} = 0.992^{12} = 0.908.$$

Because $0.908 = 1 - 0.092$, the annual decay rate is $r = 0.092$ or 9.2% which is slightly less than 12 times the monthly decay rate.

19. Extinction by Poaching. A percentage decay rate of 10% per year corresponds to a fraction decay rate of $r = 0.1$ per year. Setting $Q_0 = 1000$, the decay law is

$$Q = Q_0 \times (1 - 0.10)^t = 1000 \times 0.9^t.$$

We need to find the value of t that make $Q = 30$, the level of extinction. We can proceed by trial and error, or use logarithms. If we set $Q = 30$ in the decay law, we have

$$30 = 1000 \times 0.9^t.$$

Dividing both sides of this equation by 1000 gives us

$$0.03 = 0.9^t.$$

Taking logarithms of both sides and using the properties of logarithms (recall that $\log a^b = b \log a$), we see that

$$\log(0.03) = t \log 0.9.$$

Evaluating the logarithms and dividing both sides by $\log 0.9$, we find the that time at which the population reaches 30 is $t = 33.28$ years. After about 33 years the population will reach the critical level of 30 individuals.

21. Pesticide Decay and Your Cat.

a. With a half-life of one week, we will let t stand for time measured in weeks. The amount of pesticide on the lawn will be denoted Q and its initial value is $Q_0 = 10$ gm/sq meter. Then the decay law for the pesticide is

$$Q = Q_0 \times (1/2)^t = 10 \text{ gm/m}^2 \times (1/2)^t.$$

Notice that Q decreases by a factor of one-half each week. How long does it take for the pesticide concentration to reach 10% of its initial value or 1 gm/m²? We can follow the weekly halving of the concentration as it decreases to 5 gm/m², to 2.5 gm/m², to 1.25 gm/m² to 0.625 gm/m². We see that after 4 weeks the pesticide concentration is below 1 gm/m². Logarithms can be used to find the precise time between 3 and 4 weeks when the concentration is exactly 1 gm/m².

b. Recall that 1 m² = (100 cm)² = 10,000 cm². At a concentration of 10 gm/m², the amount of pesticide in a lunch consisting of 5 cm² of grass is

$$10 \frac{\text{gm}}{\text{meter}^2} \times \frac{1 \text{ meter}^2}{10^4 \text{ cm}^2} \times 5 \text{ cm}^2 = 0.005 \text{ grams}.$$

The pesticide on 5 cm² of grass on the first day is about 5 milligrams.

c. From part (b) we see that the pesticide level is below the maximum level of 100 milligrams on the first day. Your cat can graze as long as her diet remains below the usual 5 cm² of grass.

23. Radioactive Uranium Dating. We can follow Example 10 of the text.

a. In this case, we are given the ratio of the current to original amounts of uranium:

$$\frac{\text{current amount}}{\text{original amount}} = 85\% = 0.85.$$

We use half-life of $T_{half} = 4.5$ billion years and solve for the age t of the rock:

$$t = T_{half} \times \frac{\log_{10}\left(\dfrac{\text{current amount}}{\text{original amount}}\right)}{\log_{10}(1/2)}$$

$$= 4.5 \text{ billion years} \times \frac{\log_{10}(0.85)}{\log_{10}(1/2)}$$

$$= 1.05 \text{ billion years}.$$

The rock formed about one billion years ago.

b. We are given the ratio of the current to original amounts of uranium:

$$\frac{\text{current amount}}{\text{original amount}} = 55\% = 0.55.$$

We use half-life of $T_{half} = 4.5$ billion years and solve for the age t of the rock:

$$t = T_{half} \times \frac{\log_{10}\left(\dfrac{\text{current amount}}{\text{original amount}}\right)}{\log_{10}(1/2)}$$

$$= 4.5 \text{ billion years} \times \frac{\log_{10}(0.55)}{\log_{10}(1/2)}$$

$$= 3.88 \text{ billion years}.$$

The rock formed almost four billion years ago. Notice that the rock in part (a) is younger because it contains more of the original uranium.

25. Valium Metabolism.

a. Let's call the concentration of valium Q and as usual, Q_0 will be the initial value. We want to find the general decay law for the concentration. There are two equivalent forms of the general decay formula that we might use. Because we are given that the half-life is 36 hours, we can write immediately that

$$Q = Q_0 \times \left(\frac{1}{2}\right)^{t/36},$$

where $t = 0$ corresponds to midnight. This law insures that the value of Q decreases by a factor of 1/2 every 36 hours.

To find the concentration of valium at noon the next day, we set $t = 12$ hours with an initial value of $Q_0 = 20$ mg. The above decay law gives

$$Q = 20 \times (1/2)^{12/36} = 15.9 \text{ mg}.$$

The concentration at noon the next day is less than 16 milligrams.

b. The goal is to find the time at which Q is 10% of its initial value or 2 mg. In order to solve this problem exactly and avoid trial and error, we must use logarithms. We will use base-10 logarithms, but you can use any base you like. Starting with the growth law in the form

$$Q = 20 \times (1/2)^{t/36},$$

we first set $Q = 2$ since that is the quantity of interest. This gives us the equation

$$2 = 20 \times (1/2)^{t/36}$$

to be solved for t. Dividing both sides of this equation by 20 simplifies matters a bit:

$$0.1 = (1/2)^{t/36}$$

The fact that t appears in the exponent tells us that logarithms must be used. Let's take logarithms base-10 of both sides of the equation which gives us

$$\log_{10}(0.1) = \log_{10}((1/2)^{t/36}).$$

The crucial property of logarithms that must be used at this point is that $\log_{10}(a^b) = b \log_{10} a$. Also note that $\log_{10}(0.1) = -1$ (since $10^{-1} = 0.1$). So we have

$$-1 = (t/36)\log_{10}(1/2) = (t/36)(-0.3).$$

If we multiply both sides of this expression by 36 and divide both sides by -0.3, we find that $t = 120$ hours. This says that the drug reaches a level of 2 mg after 120 hours. You can check that if you substitute $t = 120$ into the original growth law, the result is $Q = 2$ mg.

Unit 7D Real Population Growth

Overview

In the previous units of this chapter, we have seen that exponential growth models allow only for rapid and continual growth. While this sort of growth is realistic for some populations (for example, bacteria, tumor cells, or small populations of larger animals), it cannot continue forever. Eventually lack of space or resources must limit growth.

In this short unit we look briefly at more realistic approaches to population modeling. We begin by discussing how the overall growth rate really consists of birth rates and death rates (and immigration rates). We then introduce the important concept of **carrying capacity** — the maximum number of individuals that the environment can sustain. Any realistic population model must account for the carrying capacity.

The next observation is that as a population grows, its relative growth rate cannot remain constant, as it does in an exponential growth model. The **logistic model** of growth uses a relative growth rate that actually decreases as the population grows, and reaches zero when the population reaches the carrying capacity. We present some logistic models for real population growth; then we close the unit with a discussion of estimating the carrying capacity of the Earth.

Key Words and Phrases

carrying capacity   logistic growth model   overshoot
collapse

Key Concepts and Skills

- determine the exponential growth rate from birth and death rates.
- understand the limitations of exponential growth models.
- understand the assumptions and effects of a logistic growth model.
- determine growth rates for a logistic model.
- describe overshoot and collapse models.

Solutions Unit 7D

1. Population Growth in Your Lifetime. We can form a growth law for world population by letting $t = 0$ correspond to 1996 when the population was $Q_0 = 5.8$ billion. The percentage growth rate of 1.4% per year gives a fractional growth rate of $r = 0.014$ per year. The doubling time for the population is approximately $T_{double} = 70/1.4 = 50$ years. The growth law can be written in either of two forms (where Q has units of billions of people):

$$Q = Q_0 \times (1 + r)^t = 5.8 \times (1.014)^t$$

or

$$Q = Q_0 \times 2^{t/50} = 5.8 \times 2^{t/50}.$$

These two forms must give the same results. Suppose you were born in 1970; then you will be 50 years old in 2020 which corresponds to $t = 24$. The world population according to this model will be

$$Q = 5.8 \times (1.014)^{24} = 8.10 \text{ billion}$$

or

$$Q = 5.8 \times 2^{24/50} = 8.09 \text{ billion}.$$

The two forms disagree slightly because of the approximate formula used for the doubling time.

You will be 80 years old in 2050 which corresponds to $t = 54$. The world population according to this model will be

$$Q = 5.8 \times (1.014)^{54} = 12.29 \text{ billion}.$$

You will be 100 years old in 2070 which corresponds to $t = 74$. The world population according to this model will be

$$Q = 5.8 \times (1.014)^{74} = 16.23 \text{ billion}.$$

3. Changing Population Growth Rates. We can follow Example 1 of the text to find the average growth rate over several years.

a. We are given that the population in 1900 ($t = 0$) was 1.5 billion and in 2000 ($t = 100$) is 6 billion. The growth law tells us that

$$Q = Q_0 \times (1 + r)^t$$

Now the fractional growth rate in the growth law is an unknown and all the other quantities are known. Substituting the known quantities, we have

$$6 \text{ billion} = 1.5 \text{ billion} \times (1 + r)^{100}.$$

We begin solving for r by dividing both sides by 1.5 billion:

$$4 = (1 + r)^{100}.$$

To isolate the unknown r, we take the 100th root of both sides of the equation:

$$\sqrt[100]{4} = 1 + r \quad \Rightarrow \quad 1.014 = 1 + r \quad \Rightarrow \quad r = 0.014.$$

A fractional growth rate of $r = 0.014$ per year corresponds to a percentage growth rate of 1.4% per year. An annual percentage growth rate of 1.4% per year would cause a population to grow (exponentially) from 1.5 billion to 6 billion in 100 years. This growth rate exceeds the average growth rate between 1850 and 1950.

b. We are given that the population in 1950 ($t = 0$) was 2.5 billion and in 2000 ($t = 50$) is 6 billion. The growth law tells us that

$$Q = Q_0 \times (1 + r)^t$$

The fractional growth rate, r, is an unknown and all the other quantities are known. Substituting the known quantities, we have

$$6 \text{ billion} = 2.5 \text{ billion} \times (1 + r)^{50}.$$

We begin solving for r by dividing both sides by 2.5 billion:

$$2.4 = (1 + r)^{50}.$$

To isolate the unknown r, we take the 50th root of both sides of the equation:

$$\sqrt[50]{2.4} = 1 + r \quad \Rightarrow \quad 1.018 = 1 + r \quad \Rightarrow \quad r = 0.018.$$

A fractional growth rate of $r = 0.018$ per year corresponds to a percentage growth rate of 1.8% per year. An annual percentage growth rate of 1.8% per year would cause a population to grow (exponentially) from 2.5 billion to 6 billion in 50 years. This growth rate exceeds the average growth rate between 1900 and 2000.

c. We are given that the population in 1970 ($t = 0$) was 3.7 billion and in 2000 ($t = 30$) is 6 billion. The growth law tells us that

$$Q = Q_0 \times (1 + r)^t =$$

The fractional growth rate, r, is an unknown and all the other quantities are known. Substituting the known quantities, we have

$$6 \text{ billion} = 3.7 \text{ billion} \times (1 + r)^{30}.$$

We begin solving for r by dividing both sides by 3.7 billion:

$$1.62 = (1 + r)^{30}.$$

To isolate the unknown r, we take the 30th root of both sides of the equation:

$$\sqrt[30]{1.62} = 1 + r \quad \Rightarrow \quad 1.016 = 1 + r \quad \Rightarrow \quad r = 0.016.$$

A fractional growth rate of $r = 0.016$ per year corresponds to a percentage growth rate of 1.6% per year. An annual percentage growth rate of 1.6% per year would cause a population to grow (exponentially) from 3.7 billion to 6 billion in 30 years. This growth rate is less than the average growth rate between 1950 and 2000.

d. The annual percentage growth rates over the last 100 years, the last 50 years, and the last 30 years were 1.4%, 2.2%, and 1.6%, respectively. Clearly world population accelerated during the middle part of the century and has slowed in the last part of the century.

5. Birth and Death Rates (Israel).

a. The birth rate in Israel decreased by about 25% between 1975 and 1995.

b. The death rate in Israel decreased slightly between 1975 and 1995.

c. The net growth rate is the difference between the birth and death rate. The net growth rate in 1975, 1985, and 1995 was 21.1, 16.9, and 14.7 per 1000, respectively.

d. Based on these figures, it is plausible that the population of Israel will increase, but at a slower rate in the next 20 years.

7. Birth and Death Rates (United States).

a. The birth rate in United States increased then decreased slightly between 1975 and 1995.

b. The death rate in United States was virtually constant between 1975 and 1995.

c. The net growth rate is the difference between the birth and death rate. The net growth rate in 1975, 1985, and 1995 was 5.1, 7.0, and 6.3 per 1000, respectively.

d. Based on these figures, it is plausible that the population of United States will continue to increase, perhaps at a decreasing rate, in the next 20 years.

9. Logistic Growth Rates. Following the discussion in the text and Example 3, the logistic growth rate depends on both the exponential (or base) annual growth rate, which in this case is $r = 0.03$ per year, and the carrying capacity, which is 50 million. The logistic growth is given by

$$r \times \left(1 - \frac{\text{population}}{\text{carrying capacity}}\right) = 0.03 \times \left(1 - \frac{\text{population}}{50 \text{ million}}\right).$$

When the population is 10 million, the growth rate is

$$0.03 \times \left(1 - \frac{10 \text{ million}}{50 \text{ million}}\right) = 0.024 = 2.4\%.$$

When the population is 30 million, the growth rate is

$$0.03 \times \left(1 - \frac{30 \text{ million}}{50 \text{ million}}\right) = 0.012 = 1.2\%.$$

When the population is 45 million, the growth rate is

$$0.03 \times \left(1 - \frac{45 \text{ million}}{50 \text{ million}}\right) = 0.003 = 0.3\%.$$

Notice that as the population approaches the carrying capacity, the growth rate approaches zero.

11. Logistic Carrying Capacity. The base exponential growth rate is $r = 4\%$ per year. With an actual growth rate of 3% when the population is 100,000, we can write

$$0.03 = 0.04 \times \left(1 - \frac{100,000}{\text{carrying capacity}}\right).$$

We can now solve this equation for the carrying capacity. Dividing both sides by 0.04 gives us

$$\frac{0.03}{0.04} = 1 - \frac{100,000}{\text{carrying capacity}}.$$

or

$$\frac{100,000}{\text{carrying capacity}} = 1 - \frac{0.03}{0.04} = 0.25.$$

Multiplying both sides by the carrying capacity and dividing both sides by 0.25 tells us that

$$\text{carrying capacity} = \frac{100,000}{0.25} = 400,000.$$

The carrying capacity for the population is 400,000 people.

8. LIVING WITH THE ODDS

Overview

Probability is involved in nearly every decision we make. Often it is used on a subjective or intuitive level; occasionally we try to be more precise. As you will see in this chapter, probability is one of the older and most applicable branches of mathematics. Before we are done with the chapter, we will be able to apply probability to lotteries, gambling, life insurance problems and air traffic safety. We will cover a lot of ground in this single chapter, starting with methods of counting in the Unit A, to the basics of probability in the Units B and C, to applications in Units D and E. It is a fascinating chapter, full of mathematics and real-life problems.

Unit 8A Principles of Counting

Overview

We all know how to count *individual* objects or people, but there are many other types of counting problems that arise when we want to count *groups* of objects or people. That is the subject of this chapter. There are four basic counting methods presented in this chapter:

- selections from two or more groups (multiplication principle),
- arrangements with repetition,
- permutations, and
- combinations.

In this unit, we will learn not only how to apply these methods, but equally important, how to determine *when* to use each method.

Selection from two or more groups occurs when we have several different groups of objects and we need to select one item from each group. For example, in a restaurant, you might have four choices for an appetizer (first group), six choices for a main course (second group), and five choices for a dessert (third choice). How many different meals could you select?

Problems of this sort can be solved using a **table**, a **tree**, or (the easiest of all) the **multiplication principle**. You will see that the total number arrangements of items from two or more groups is

(number of items in first group) × (number of items in second group) × (number of items in third group) × ···· × (number of items in last group).

In the above example, there would be $4 \times 6 \times 5 = 120$ different possible meals that you could order.

Arrangements with repetition occur when you select items from a single group and items may be used more than once. Often the items in problems of this kind are letters or numerals. For example, how many different three-digit area codes can be formed from the numerals 0 through 9? There are 10 ways to choose the first numeral and, since repetition is allowed, there are 10 ways to choose the second numeral, and 10 ways to choose the third numeral. This amounts to a total of $10 \times 10 \times 10 = 1000$ different three-digit area codes. The general rule is that

> if we make r selections (for example, the three digits of the area code) from a group of n items (for example, the numerals 0–9), n^r different arrangements are possible.

Permutations also involve selecting items from a single group with one important difference — repetition is not allowed. Furthermore, in counting permutations, the order of the arrangements matters; that is, we count ABCD and DCBA as two different arrangements. To summarize, permutations require

- selection from a single group,
- repetition is not allowed, and
- order matters.

In the unit, we work slowly through several examples towards the general formula for counting permutations. We need a bit of mathematical notation along the way, so we introduce the **factorial** function.

$$n! = n \times (n-1) \times (n-2) \times \ldots \times 2 \times 1.$$

Your temptation might be to resist learning and using factorials, but they will make your life much easier! Just a little practice is all it takes.

With factorials in hand, we can write a general formula for counting permutations.

> If we make r selections from a group of n items, the number of possible permutations is
>
> $$_nP_r = \frac{n!}{(n-r)!},$$
>
> where $_nP_r$ is read as "the number of permutations of n items taken r at a time."

Combinations are much like permutations with one difference — order does not matter; that is, we count ABCD and DCBA as the *same* arrangement. The requirements for combinations are

- selection from a single group,
- repetition is not allowed, and
- order does *not* matter.

The number of combinations of *n* objects taken *r* at a time is the same as the number of permutations, except that we have to correct for the overcounting that the permutation formula does (since permutations care about order and combinations don't). This subtle point is explained in the text. The result is the combinations formula.

If we make *r* selections from a group of *n* items, the number of possible *combinations*, in which order does not matter, is

$$_nC_r = \frac{_nP_r}{r!} = \frac{n!}{(n-r)! \times r!}$$

where $_nC_r$ is read as "the number of combinations of *n* items taken *r* at a time."

Make sure that you keep Table 8.1 nearby; it summarizes the four counting methods of this unit. As said earlier, they key is knowing not only *how* to use a particular method, but *when* to use it. There are plenty of practice problems at the end of the unit; you should work as many as possible. Another hint is to learn the capabilities of your calculator. You will need a calculator that at least does exponentiation (n^r) Many calculators compute factorials directly with a special factorial key. Some calculators have keys for permutations and combination. You will save yourself considerable work if you let your calculator do as much work as possible.

Key Words and Phrases

selection from two or more groups	multiplication principle	arrangements with repetition
permutation	combination	

Key Concepts and Skills

- describe the four different counting methods discussed in the unit.
- determine which of the four methods applies to a given counting problem.
- use factorials confidently.
- apply each of the four methods on appropriate problems.

Unit 8A Solution

Choosing from Two Groups.

1. From the table and tree below, we see that there are 24 different choices of cars. By the multiplication principle, there are 8×3 choices.

	Sedan	Station Wagon	Hatchback
Color 1	Car #1	Car #2	Car #3
Color 2	Car #4	Car #5	Car #6
Color 3	Car #7	Car #8	Car #9
Color 4	Car #10	Car #11	Car #12
Color 5	Car #13	Car #14	Car #15
Color 6	Car #16	Car #17	Car #18
Color 7	Car #19	Car #20	Car #21
Color 8	Car #22	Car #23	Car #24

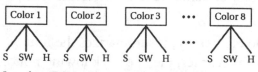

S = sedan   SW = station wagon   H = hatchback

3. Selections are made from two different groups (colors and patterns). Using a table we can represent the arrangements of colors and patterns as follows (where C1 means Color #1, P1 means Pattern #1, etc.).

	C1	C2	C3	C4	C5	C6	C7	C8
P1	1	2	3	4	5	6	7	8
P2	9	10	11	12	13	14	15	16
P3	17	18	19	20	21	22	23	24
P4	25	26	27	28	29	30	31	32

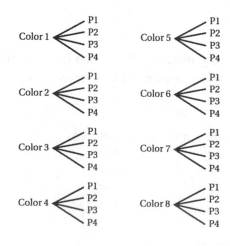

The simplest way to find the total number of arrangements is to use the multiplication principle. Because there are eight choices for color and four choices for pattern, there are $8 \times 4 = 32$ possible wallpaper selections.

Choosing from More Than Two Groups.

5. There are 8 choices of skis, six choices of bindings, and seven choices of boots that can be arranged in any ski/binding/boot package. By the multiplication principle, there are $8 \times 6 \times 7 = 336$ different packages.

7. In this case, we are making selections from five different groups: humanities courses (of which there are four to choose), sociology courses (of which there are three to choose), science courses (of which there are five to choose), math courses (of which there are two to choose), and music courses (of which there are three to choose). By the multiplication principle, there are $4 \times 3 \times 5 \times 2 \times 3 = 360$ different schedules that can be made using one course from each of the five groups.

9. Arrangements with Repetition.

a. There are 10 numerals to choose for each of the four positions and numerals can be used more than once. So we are counting arrangements with repetition. There are 10 numerals for the first position, 10 numerals for the second position, 10 numerals for the third position, and 10 numerals for the fourth position, for a total of $10 \times 10 \times 10 \times 10 = 10,000$ different house numbers. Note that if we excluded zero as the first digit, there would be $9 \times 10 \times 10 \times 10 = 9,000$ different house numbers.

b. There are seven different notes that can be used and they are to be placed (with repetition allowed) in the 10 different positions of the "tune." There are

$$7 \times 7 \times \ldots \times 7 \ (10 \text{ times}) = 7^{10} = 282,475,249$$

different "tunes" that can be composed.

c. There are 26 letters that can be placed (with repetition allowed) in each of the five positions of the word. This means that there are

$$26 \times 26 \times 26 \times 26 \times 26 = 26^5 = 11,881,376$$

different words that can be created.

d. The word "combination" might be misleading in this problem. We are counting *lock combinations*, but the arrangements are not combinations in the mathematical sense; they are arrangements with repetition. There are 8 different ways to set the first dial (A through H). Since repetition *is* allowed, there are also 8 different ways to set the second dial and 8 different ways to set the third dial. Thus, $8 \times 8 \times 8 = 512$ different arrangements are possible.

11. Factorial Practice.

a. $7! = 7 \times 6 \times 5 \times 4 \times 3 \times 2 \times 1 = 5040.$

b. $\dfrac{7!}{3!} = \dfrac{7 \times 6 \times 5 \times 4 \times 3 \times 2 \times 1}{3 \times 2 \times 1} = 840.$

c. $\dfrac{7!}{4!3!} = \dfrac{7 \times 6 \times 5 \times 4 \times 3 \times 2 \times 1}{(4 \times 3 \times 2 \times 1)(3 \times 2 \times 1)} = 35.$

d. $\dfrac{12!}{5!} = 12 \times 11 \times 10 \times 9 \times 8 \times 7 \times 6 = 3{,}991{,}680.$

e. $\dfrac{12!}{8!(12-8)!} = \dfrac{12!}{8!4!} = \dfrac{12 \times 11 \times 10 \times 9}{4 \times 3 \times 2 \times 1} = 495.$

f. $\dfrac{21!}{20!} = \dfrac{21 \times 20 \times 19 \times \cdots \times 3 \times 2 \times 1}{20 \times 19 \times \cdots \times 3 \times 2 \times 1} = 21.$

13. Permutations. These questions deal with permutations because we are making selections from a single group, repetitions are not allowed (once a person has been chosen for a committee or team, s/he cannot be chosen again), and order does matter.

a. First note why this is a permutation problem: We are making selections from a single group (the group of 12 photographs), repetitions are not allowed, and order matters. The number of different arrangements is the number of permutations of 12 objects taken 5 at a time or

$$_{12}P_5 = \frac{12!}{(12-5)!} = \frac{12!}{7!} = 12 \times 11 \times 10 \times 9 \times 8$$
$$= 95{,}040 \text{ arrangements.}$$

b. In making passwords, order *does* matter (ABCD is different from DBCA). Since repetition is not allowed, we must count permutations. The number of ways that six letters can be arranged in groups of five is

$$_6P_5 = \frac{6!}{(6-5)!} = \frac{6!}{1!} = 6 \times 5 \times 4 \times 3 \times 2 = 720.$$

A total of 720 different passwords can be formed.

c. Because all of the acts must be scheduled, we must arrange 12 people in groups of 12. Order does matter and repetition is not allowed. The number of different schedules is

$$_{12}P_{12} = \frac{12!}{(12-12)!} = \frac{12!}{0!} = 12! = 479{,}001{,}600.$$

Recall the definition that $0! = 1$. We see that almost half a billion different schedules can be made.

d. Because a card may not be used twice and order matters, we must count the permutations of 52 cards taken 10 at a time. The number of card sequences is

$$_{52}P_{10} = \frac{52!}{(52-10)!} = \frac{52!}{42!} = 5.74 \times 10^{16}.$$

This result is not exact, but approximately 57 quadrillion different card sequences are possible.

15. Combinations. These questions deal with combinations because we are making selections from a single group, repetitions are not allowed (once a person has been chosen for a committee or team, s/he cannot be chosen again), and order does not matter.

a. The number of five-person subcommittees that can be formed from ten people is

$$_{10}C_5 = \frac{10!}{5!(10-5)!} = \frac{10!}{5!5!} = \frac{10 \times 9 \times 8 \times 7 \times 6}{5 \times 4 \times 3 \times 2 \times 1} = 252.$$

There are 252 different four-person committees that can be chosen from a 12-member group.

b. Card hands usually involve combinations since order does not matter. If we deal four-card hands from a 40-card deck, then the number of different hands that can be dealt is

$$_{40}C_4 = \frac{40!}{4!(40-4)!} = \frac{40!}{4!36!} = \frac{40 \times 39 \times 38 \times 37}{4 \times 3 \times 2 \times 1} = 91{,}390.$$

c. We assume that order is not important so we are counting combinations. The number of ways that six disks can be selected from a group of ten disks is

$$_{10}C_6 = \frac{10!}{6!(10-6)!} = \frac{10!}{6!4!} = \frac{10 \times 9 \times 8 \times 7}{4 \times 3 \times 2 \times 1} = 210.$$

The ten disks can be arranged in 210 different groups of six if order does not matter and no repetition is allowed.

d. The task is to find the number of combinations of 20 people taken 8 at a time. Since we are counting combinations, the number of teams is

$$_{20}C_8 = \frac{20!}{8!(20-8)!} = \frac{20!}{8!12!}$$
$$= \frac{20 \times 19 \times 18 \times 17 \times 16 \times 15 \times 14 \times 13}{8 \times 7 \times 6 \times 5 \times 4 \times 3 \times 2 \times 1} = 125{,}970.$$

A total of 125,970 different eight-person track squads can be chosen.

P 17. Permutation and Combination Practice.

a. $_9C_3 = \dfrac{9!}{6!3!} = \dfrac{9 \times 8 \times 7}{3 \times 2} = 84$ is the number of three-card arrangements that can be made from nine cards.

b. $_9P_3 = \dfrac{9!}{6!} = 9 \times 8 \times 7 = 504$ is the number of ways of arranging three books on a shelf, choosing from a group of nine books.

c. $_4P_2 = \dfrac{4!}{2!} = 4 \times 3 = 12$ is the number of ways to choose a president and vice-president from a four-person board of directors.

d. $_4C_2 = \dfrac{4!}{2!2!} = \dfrac{4 \times 3}{2} = 6$ is the number of two-person subcommittees that can be formed from four people.

e. $_{12}C_8 = \dfrac{12!}{4!8!} = \dfrac{12 \times 11 \times 10 \times 9}{4 \times 3 \times 2} = 495$ is the number of eight-topping pizzas that can be made with 12 toppings to choose from.

f. $_{12}P_8 = \dfrac{12!}{4!} = 12 \times 11 \times 10 \times 9 \times 8 \times 7 \times 6 \times 5 = 19{,}958{,}400$

is the number of eight-team rankings that can be made of 12 teams.

19. Manager's Job. In choosing a lineup, order does matter, so we must count permutations. The number of nine-person batting orders that can be formed from 25 people is

$$_{25}P_9 = \dfrac{25!}{(25-9)!} = \dfrac{25!}{16!} = 7.41 \times 10^{11}.$$

There are over 700 billion lineups that can be made! Notice that when the answers get too large it is easiest (and permissible) to round them and use scientific notation. If a different lineup was used every day, it would take 7.41×10^{11} days or

$$7.41 \times 10^{11} \text{ days} \times \dfrac{1 \text{ yr}}{365 \text{ days}} = 2 \times 10^9 \text{ years}.$$

It would take two billion years to try every lineup!

21. Making a Toast. When people toast, they combine themselves in pairs. In counting the pairwise arrangements in a toast, order does not matter (you would not count Kate and Bill as one toast and Bill and Kate as another toast). Therefore, we must count the number of combinations of eight people taken two at a time. The number of toasts is $_8C_2 = 28$.

23. House Numbers. The house numbers on Sierra Drive must have the form 8XY where Y is an odd number (since the entire number must be odd) and X can be any number. There are 10 choices for X (0 through 9) and 5 choices for Y (1, 3, 5, 7, 9). By the multiplication principle, there are $10 \times 5 = 50$ different three-digit addresses possible.

25. Passwords of Symmetric Letters. We have 11 characters to arrange in groups of 6. If repetition is allowed there are $11^6 = 1{,}771{,}561$ passwords possible. If repetition is not allowed and order matters, then $_{11}P_6 = 332{,}640$ passwords are possible.

27. Telephone Numbers.

a. Forming telephone numbers is an example of arrangements with repetition since each digit can be used more than once in a telephone number. There are eight choices for the first digit (because 0 and 1 are excluded) and ten choices for the remaining six digits. The number of possible seven-digit phone numbers is

$$8 \times 10 \times 10 \times 10 \times 10 \times 10 \times 10 = 8{,}000{,}000.$$

Therefore, two million people can be served by a set of seven-digit phone numbers.

b. Within a single exchange there are four digits to be chosen, so there are $10^4 = 10{,}000$ possible telephone numbers within a single exchange. This is probably not enough numbers to serve 80,000 people since there are fewer than eight people per telephone number in most cities. If we assume that on average four people share a single telephone number, then we would need 20,000 phone numbers in a city of 80,000 people, which would require two exchanges. The actual number of possible phone numbers is lower than the above figures since some digits (for example, a leading zero) are excluded from exchanges and area codes. For example, if a leading zero were excluded from the exchange, then we could make 9×10^6 or nine million (instead of ten million) different phone numbers within a single area code.

29. ZIP Codes.

a. If we assume that all ten digits (0 through 9) can be used in all five positions of the zip code, then 10^5 or 100,000 zip codes are possible.

b. If these 100,000 zip codes were spread out evenly over all 270 million people, then on average there would be $(2.7 \times 10^8 \text{ people})/(10^5 \text{ zip codes}) = 2700$ people per zip code. In practice, many more people are assigned to a given zip code. Not all of the zip codes are used because of some restrictions.

c. If we assume that all ten digits (0 through 9) can be used in all *nine* positions of a zip code, then 10^9 or one billion zip codes are possible. Nine digits increases the number of possible zip codes by a factor of 10^4 or 10,000. With nine-digit zip codes everyone in the United States could have a personal zip code.

31. Tournaments.

a. Since the order does not matter in pairing teams (Team A plays Team B is the same as Team B plays Team A), we are counting combinations. The number of ways that 16 teams can be grouped two at a time is $_{16}C_2 = 120$.

b. With eight teams there are $_8C_2 = 28$ ways to arrange them two at a time.

c. With four teams there are $_4C_2 = 6$ ways to arrange them two at a time.

33. Shuffling Cards.

a. Since we are arranging the entire deck, order *does* matter in counting groupings of the cards. The number of different ways to arrange all 52 cards of a deck is

$$_{52}P_{52} = 52 \times 51 \times 50 \times \ldots \times 3 \times 2 \times 1 = 8.1 \times 10^{67}.$$

b. The question asks that we calculate the number of years in 8.1×10^{67} minutes. Using familiar unit conversions, we find that this length of time in years is

$$8.1 \times 10^{67} \text{ min} \times \frac{1 \text{ hr}}{60 \text{ min}} \times \frac{1 \text{ day}}{24 \text{ hr}} \times \frac{1 \text{ yyr}}{365 \text{ day}} = 1.5 \times 10^{62} \text{ yr.}$$

The time required to make all of the arrangements of a deck of cards far exceeds the expected lifetime of the Sun.

Unit 8B Fundamentals of Probability

Overview

The work we did with counting methods in the previous unit was not in vain. Now we can put it to use to compute probabilities. We all have an intuitive sense of what a probability is and that meaning is useful. A probability is a number between zero and one. If the probability of an event is zero, then it cannot happen; if the probability of an event is one, then it is certain to happen; and if the probability is somewhere is in between zero and one, it gives a measure of how likely the event is to happen.

We can approach probabilities in three different ways:

- an *a priori* **probability** uses mathematical methods to find the probability of an event occurring.
- an **empirical probability** is determined with experiments or by data collection .
- a **subjective probability** is based on intuition.

This unit will focus on *a priori* probability calculations; it will mention empirical probabilities, and we will leave subjective probabilities for another course!

If you wanted to find the probability that a fair dice will show a 5 when rolled once, you might reason that there are six possible outcomes of a single roll and a success (rolling a 5) is one of those six outcomes. So you might reason that the probability of rolling a 5 is 1/6. You would be correct! This thinking is the basic procedure for computing *a priori* probabilities:

Step 1:   Count the total number of possible outcomes of an event.

Step 2:   Count the number of outcomes that represent **success** — that is, the number of outcomes that represent the sought after result.

Step 3:   Determine the probability of success by dividing the number of successes by the total number of possible outcomes:

$$\text{probability of success} = \frac{\text{number of outcomes that represent success}}{\text{total number of possible outcomes}}$$

This procedure can be used to compute the probability of many events as you will see in the examples of the text. You will also see how the counting methods of the previous unit are used in steps 1 and 2 of this procedure.

Another important fact that will be used repeatedly is that if the probability of an event occurring is p, then the probability that it does not occur is $1 - p$. For example, the probability of rolling a fair die and getting a five is 1/6; therefore the probability of rolling *anything but* a five is $1 - 1/6 = 5/6$.

Suppose you flip three coins and are interested in *all* of the outcomes. In this case, there are four possible outcomes: three heads, two heads and one tail, one head and two tails, and three tails. We will show how to find the probability of all of the outcomes. The result is called a **probability distribution** which is just another way of saying the probabilities of *all* possible outcomes.

There are many situations in which it's impossible to determine probabilities *a priori*. In these situations, we can't use mathematical methods because we have only records or data to work with. For example, we often talk about a 100-year flood, which means that based on historical records, a flood of this magnitude occurs once every 100 years. So the probability of such a flood happening in any given year is 1/100. We discuss empirical probabilities because they are used so often in practice.

Finally, we take a few pages to clarify the confusion that arises over the use of the word *odds*. The odds of an event A happening are given by

$$\text{odds}(A) = \frac{\text{probability of A}}{\text{probability of not A}} = \frac{P(A)}{P(\text{not } A)}$$

For example, the odds of tossing a fair coin and getting a head are 1, often said *1 to 1*. Unfortunately, the term odds has an even slightly different meaning when used in gambling. There, it is the amount a winning bet will pay for every dollar that you bet. Needless to say, care must be used when dealing with odds (and gambling!).

Key Words and Phrases

a priori probability	empirical probability	subjective probability
probability	odds	
distribution		

Key Concepts and Skills

- distinguish between *a priori* probability, empirical probability, and subjective probability.
- use the three-step process to determine simple *a priori* probabilities.
- find probability distributions for coin and dice experiments.
- determine empirical probabilities from data or historical records.
- find the probability of an event *not* occurring.
- find the odds of an event from its probability.

Unit 8B Solutions

5. In drawing a card from a deck there are 52 possible outcomes and 4 ways that the card can be a 2 (a success). Thus, the probability of drawing a 2 is P(2) = 4/52 = 1/13.

7. In drawing a card from a deck there are 52 possible outcomes and 26 ways that the card can be a red card (a success). Thus, the probability of drawing a red card is P(red) = 26/52 = 1/2.

9. In drawing a card from a deck there are 52 possible outcomes and 39 ways that the card can fail to be a spade (a success). Thus, the probability of drawing a card besides a spade is P(no spade) = 39/52 = 3/4 which is one minus the probability of drawing a spade.

11. Bag of Marbles. There are 17 possible outcomes (because there are 17 marbles in the bag).

a. A success is drawing a red marble, of which there are two. Thus, the probability of drawing a red marble is 2/17.

b. A success is drawing a blue marble, of which there are five. Thus, the probability of drawing a blue marble is 5/17.

c. A success is drawing a white marble, of which there are ten. Thus, the probability of drawing a white marble is 10/17.

d. A success is drawing a blue or red marble (not a white marble). There are seven blue or red marbles. Thus, the probability of not drawing a white marble is 7/17 which is one minus the probability of drawing a white marble.

13. Probability of Non-Events. Recall that the probability of an event *not* happening is one minus the probability of the event happening.

a. The probability of getting two heads when flipping two coins is 1/4: P(2 heads) = 1/4. Thus P(not 2 heads) = 1 − P(2 heads) = 3/4.

b. The probability of getting an odd number on the roll of a fair die is P(odd) = 3/6 = 1/2 (because there are 3 odd numbers and 6 outcomes). Thus P(not odd) = 1 −P(odd) = 1 − 1/2 = 1/2.

c. Because there are 12 face cards in a deck of 52 cards, P(face card) = 12/52 = 3/13. Thus P(not face card) = 1 − 3/13 = 10/13.

d. There are 6 double numbers among the 36 outcomes when two dice are rolled, so P(double) = 6/36 = 1/6. Thus P(not double) = 1 − 1/6 = 5/6.

15. Four Coin Probability Distribution.

a. Here is the table showing all possible outcomes of tossing four coins.

Coin #1	Coin #2	Coin #3	Coin #4
H	H	H	H
H	H	H	T
H	H	T	H
H	H	T	T
H	T	H	H
H	T	H	T
H	T	T	H
H	T	T	T
T	H	H	H
T	H	H	T
T	H	T	II
T	H	T	T
T	T	H	H
T	T	H	T
T	T	T	H
T	T	T	T

b. It helps to consolidate the results of the above table and show the number of occurrences of each different result.

Result	# Occurrences	Probability
4H	1	1/16
3H, 1T	4	1/4
2H, 2T	6	3/8
1H, 3T	4	1/4
4T	1	1/16

c. We see from the table that the probability of tossing three heads and one tail is P(3H,1T) = 1/4.

d. There are 15 ways of tossing anything but four tails, so the probability of this event is 15/16.

17. Two Dice Probabilities. We can refer to the table in Example 5 of the text.

a. Because there are 3 ways that the two dice can have a sum of 4 and there are 36 possible outcomes, P(sum of 4) = 3/36 = 1/12.

b. Using part (a), P(not sum of 4) = 1 – P(sum of 4) = 1 – 1/12 = 11/12.

c. There are two ways to roll a 5 on one die and a 2 on the other die. Therefore P(2 and 5) = 2/36 = 1/18.

d. The table shows that the number of times that one (but not both) of the dice is a 3 is 10. The total number of outcomes is 36, so P(one 3) = 10/36 = 5/18. If a double three is included, then P(one 3) = 11/36.

e. Imagine that you play this game 36 times (although you can pick any number of games). In 36 games you would expect (on average) to roll a sum of 6 and win 5 times. You would expect to roll a sum other than 6 and lose 31 times. Therefore, you would expect to win a total of 5 × $10 = $50 and lose a total of 31 × $1 = $31. Thus, in the long run, you would expect to win $50 for every $31 you lose; or win $50/31 = $1.61 for every $1 you lose. This game is probably worth playing.

21. Computing the Odds. Recall that the definition of odds given in the text is that the odds of an event A are P(A)/P(not A) or P(A) to P(not A). Use the terms *odds* with care; it can lead to confusion!

a. The probability of rolling a 1 or 2 is 1/3 and the probability of rolling anything but a 1 or 2 is 2/3. Therefore the odds of rolling a 1 or 2 are (1/3)/ (2/3) = 1/2. We would say that the odds are 1 to 2 (less than 50-50).

b. In flipping two fair coins, P(2 tails) = 1/4 and P(not 2 tails) = 3/4. The odds of getting two tails are (1/4)/(3/4) = 1/3 or 1 to 3.

c. In drawing a single card from a standard deck, P(spade) = 1/4 and P(not spade) = 3/4. The odds of drawing a spade are (1/4)/(3/4) = 1/3 or 1 to 3

d. In drawing a single card from a standard deck, P(face card) = 12/52 = 3/13 and P(not face card) = 10/13. The odds of drawing a face card are (3/13)/(10/13) = 3/10 or 3 to 10.

23. Gambling Odds. Recall that the use of odds in betting is slightly different than the use of odds for expressing a probability.

a. Odds of 3 to 4 mean that for every $4 you bet you gain $3 if you win. If you bet $20 (5 bets of $4), then you gain 5 × $3 = $15.

b. Odd of 5 to 4 mean that for every $4 you bet, you gain $5 if you win. Thus if you bet $20 (5 bets of $4) and win, you receive 5 ×$5 = $25.

25. Lotto Psychology.

a. Arrangements in a lottery in which balls are drawn at random are combinations. The number of combinations of 42 numbered balls selected 6 at a time is

$$_{42}C_6 = \frac{42!}{(42-6)!6!} = \frac{42!}{36!6!} = \frac{42 \times 41 \times 40 \times 39 \times 38 \times 37}{6 \times 5 \times 4 \times 3 \times 2 \times 1}$$

$$= \frac{3,776,965,920}{720} = 5,245,786.$$

We see that there are over five million different lottery combinations.

b. If you choose one lottery number, your chances of choosing the winning number are 1 in 5,245,786.

c. If the lottery is not rigged, every number has the same chance of being drawn. There are no special numbers such as 1,2,3,4,5,6.

Unit 8C Combining Probabilities

Overview

In the previous unit, we studied methods for finding the probability of individual events occurring. However, many interesting situations actually consist of multiple events. For example, what is the probability of tossing ten heads in a row with a fair coin? Or what is the probability of drawing a jack or a heart from a standard deck of cards? We will answer these and many other practical questions in this unit.

The unit presents five different techniques that involve multiple events. As always, it's important to know *how* to apply these techniques and *when* to apply them. The five situations covered by these methods are:

- joint independent events,

- joint dependent events,
- either/or mutually exclusive events,
- either/or non-mutually exclusive events, and
- *at least once* events.

The term **joint probability** refers to two or more events *all* happening. For example, for example what is the probability of rolling two fair dice and seeing a six on *both* dice (a six *and* a six)? Or what is the probability of drawing four cards from a standard deck and getting an ace each time? The important distinction is whether the events in question are **independent** or **dependent**. If one event does not affect the others (for example, in rolling two dice, the outcome of one die does not affect the outcome of other die), then we have a joint probability for independent events. The rule that applies for two independent events is

$$P(A \ and \ B) = P(A) \times P(B).$$

This principle can be extended to any number of independent events. For example, the joint probability of A, B, and a third independent event C is

$$P(A \ and \ B \ and \ C) = P(A) \times P(B) \times P(C).$$

If the events are not independent, then a bit more care is needed. For example, if you draw two cards from a deck, but don't replace the first card before drawing the second, then the outcome of the second card depends on the outcome of the first card. The rule that applies in this case for two dependent events is

$$P(A \ and \ B) = P(A) \times P(B \ given \ A).$$

With **either/or events** we are interested in the probability of *either* event A *or* event B occurring. The important distinction is whether the events in question are mutually exclusive. If the occurrence of the first event prevents the occurrence of the second event, then the events are **mutually exclusive**. For example, drawing a heart and drawing a diamond from a deck of cards are mutually exclusive events because if one event occurs (say, drawing a heart), then the other event (drawing a diamond) cannot occur. The rule that applies in this case with two events is

$$P(A \ or \ B) = P(A) + P(B).$$

This principle can be extended to any number of mutually exclusive events. For example, the probability that either event A, event B, or event C occurs is

$$P(A \ or \ B \ or \ C) = P(A) + P(B) + P(C).$$

If the events are **non-mutually exclusive** (for example, going into a room of people and meeting a woman *or* a Democrat), then a modification must be used. The rule that applies in this case is

$$P(A \ or \ B) = P(A) + P(B) - P(A \ and \ B).$$

In general, for joint probabilities we multiply probabilities and for either/or probabilities we multiply probabilities. However, modifications must be made in the cases of dependent or non-mutually exclusive events.

Finally, a very important situation in one in which we ask about the probability of an event happening *at least once*. For example, if you buy ten lottery tickets, what is the probability of *at least* one ticket being a winner? If you roll a die ten times, what is the probability of rolling *at least* one six? The rule that tells you the probability of an event A occurring at least once in n trials is

$$P(A \text{ at least once in } n \text{ trials}) = 1 - P(A \text{ does not occur in } n \text{ trials})$$

$$= 1 - P(not \ A)^n.$$

The unit has explanations of these rules and many examples to show how they are used. Be sure you refer to Table 8.2 for a summary of the rules. And be sure you also give yourself plenty of time to practice!

Key Words and Phrases

joint probability	independent events	dependent events
either/or probability	mutually exclusive	non-mutually exclusive
at least once rule		

Key Concepts and Skills

- distinguish between a joint probability and an either/or probability.
- distinguish between independent events and dependent events .
- distinguish between mutually exclusive events and non-mutually exclusive events.
- identify *at least once* situations.
- know when and how to apply the five probability rules given in the unit.

Solutions Unit 8C

1. Card Probabilities with Replacement. Recall that drawing cards with replacement produces independent events.

a. The probability of drawing an ace on the first draw is P(A) = 4/52 = 1/13 = 0.077 because there are four ways to draw an ace from a deck with 52 cards.

b. The probability of drawing two aces on the first two draws (with replacement) is P(AA) = P(A) × P(A) = 1/13 × 1/13 = 0.0059.

c. Similarly, the probability of drawing four aces on the first four draws (with replacement) is P(AAAA) = P(A) × P(A) × P(A) × P(A) = (P(A))^4 = (1/13)^4 = 0.000035.

d. Similarly, the probability of drawing five aces on the first five draws is P(AAAAA) = (P(A))^5 = (1/13)^5 = 0.000003.

3. Joint Probabilities.

a. The two dice are independent and the probability of rolling a 6 on either die is 1/6. Therefore, the probability of rolling a 6 on both dice is P(6,6) = 1/6 × 1/6 = 1/36 = 0.028.

b. Each coin toss is independent and the probability of a head on a single toss is 1/2. So the probability of six consecutive heads is P(6H) = (1/2)^6 = 1/64 = 0.016.

c. The outcomes of lottery tickets are independent. So with a 1/10 probability of winning with a single ticket, the probability of winning on three tickets in a row is P(3 wins) = (1/10)^3 = 1/1000 = 0.001.

d. Drawing cards with replacement and shuffling gives independent events. The probability drawing an ace is P(A) = 1/13 and the probability drawing a king is P(K) = 1/13. Therefore the probability of drawing an ace on the first draw and a king on the second draw is P(AK) = 1/13 × 1/13 = 0.0059.

5. Drawing Candy. The probability of drawing a chocolate on the first try is 12/20 = 3/5 = 0.6. So the probability of choosing a chocolate twice in a row *with replacement* is 3/5 × 3/5 = 9/25 = 0.36. However, if the first piece of candy is not replaced, then the outcome of the second draw depends on the outcome of the first. If a chocolate is drawn the first time, this leaves 11 chocolates in the bag out of a total of 19 candies. So the probability of choosing a chocolate on the second draw given that a chocolate was chosen on the first draw is 11/19. So we have P(two chocolates) = P(chocolate on first draw) × P(chocolate on second draw given a chocolate on first draw) = 3/5 × 11/19 = 0.35 — slightly less than the replacement with replacement.

7. Card Probabilities Without Replacement. Remember that card drawing without replacement produces dependent events.

a. As we have seen, the probability of drawing an ace on the first draw is P(A) = 1/13.

b. Note that P(ace given an ace on the first draw) = 3/51 (because there are three aces and 51 cards left in the deck after drawing an ace). Therefore, P(AA) = 1/13 × 3/51 = 3/663 = 0.0045.

c. Each time an ace is drawn from the deck, the number of aces decreases by one and the number of cards in the deck decreases by one. Therefore,

$$P(AAAA) = \frac{4}{52} \times \frac{3}{51} \times \frac{2}{50} \times \frac{1}{49} = 0.000004.$$

The chances of drawing four aces in a row are about four in a million.

d. Each time a card is drawn the number of cards in the deck decreases by one. On the first draw, P(ace of spades) = 1/52. Then without replacement, P(king of spades given an ace of spades on the first draw) = 1/51. Similarly, P(queen of spades given an ace of spades on the first draw and a king of spades on the second draw) = 1/50. Continuing in this way we find the probability of drawing ace, king, queen, jack, and ten of spades is

$$P = \frac{1}{52} \times \frac{1}{51} \times \frac{1}{50} \times \frac{1}{49} \times \frac{1}{48} = 3 \times 10^{-9}.$$

9. Either/Or Probabilities.

a. Drawing a 2 and drawing a 4 from a deck of 52 cards are mutually exclusive events and each event has a probability of 1/13. Therefore, the probability of drawing either a 2 or a 4 is P(2 or 4) = P(2) + P(4) = 1/13 + 1/13 = 2/13 = 0.15.

b. Rolling a 1 and rolling a 2 are mutually exclusive events and each event has a probability of 1/6. Therefore, the probability of rolling either a 1 or a 2 is P(1 or 2) = P(1) + P(2) = 1/6 + 1/6 = 1/3.

c. Rolling a sum of 7 and rolling a sum of 8 are mutually exclusive events. As we saw in Unit 8B, P(sum of 7) = 1/6 and P(sum of 8) = 5/36. Therefore, the probability of rolling a sum of either 7 or 8 is 1/6 + 5/36 = 11/36 = 0.31.

d. In this problem, we are dealing with non-mutually-exclusive events since choosing a woman does not exclude choosing a Democrat. We can reason in two ways here. (i) The only way *not* to choose a woman or a Democrat is to choose a Republican man. The probability of choosing a Republican man is 1/4, so the chance of *not* choosing a woman *or* a Democrat is 1/4. This means the probability of choosing a woman or a Democrat is 3/4.

Alternatively, we can use the formula for non-mutually-exclusive events which give us

P(woman *or* Democrat) = P(woman) + P(Democrat) –
    P(woman *and* Democrat) = 1/2 + 1/2 – 1/4 = 3/4.

Either way, the probability of choosing a woman or a Democrat from the group is 3/4 or 0.75.

e. Drawing a king and drawing a heart are not mutually exclusive events. Therefore

P(king or heart) = P(king) + P(heart) – P(king and heart)
        = 1/13 + 1/4 – 1/52 = 16/52 = 4/13.

f. Drawing a king or a diamond are not mutually exclusive events, nor are drawing an ace or a diamond. Drawing an ace or a king are mutually exclusive. So we need to be careful not to count the ace of diamonds and the king of diamonds twice. Using the formula for non-mutually-exclusive events gives us

P(ace *or* king *or* diamond) = P(ace) + P(king) +
    P(diamond) – P(ace *and* diamond) – P(king *and*
    diamond) = 1/13 + 1/13 + 1/4 – 1/52 – 1/52 = 19/52.

11. Game Spinner. Note that each of the number sectors has the same probability of occurring, namely 1/6. Each of the color sectors has the same probability of occurring, namely 1/2.

a. P(6) = 1/6.

b. P(white) = 1/2.

c. Spinning a 1 and spinning a 3 are mutually exclusive events, so P(1 or 3) = 1/6 + 1/6 = 1/3.

d. Spinning a 1 and spinning a blue are non-mutually exclusive events, so P(1 or blue) = P(1) + P(blue) – P(1 and blue) = 1/6 + 1/2 – 1/6 = 1/2.

e. Spinning a 1 and spinning a white are mutually exclusive events (because the number 1 sector is colored blue), so P(1 or white) = P(1) + P(white) =   1/6 + 1/2 = 2/3.

13. At Least Once Rule. Recall that the "at least once" rule says that

$$P(A \text{ at least once in } n \text{ trials}) = 1 - P(\text{not } A)^n,$$

provided the trials are independent.

a. The probability of rolling a 6 is P(6) = 1/6. The probability of not rolling a 6 is P(no 6) = 1 – P(6) = 5/6. The probability of not rolling a 6 in 5 rolls (independent events) is

P(no 6 in 5 rolls) = P(no 6)^5 = (5/6)^5 = 0.40.

Therefore the probability of rolling at least one 6 in 5 rolls is P(at least one 6) = 1 – 0.40 = 0.60.

b. The probability of flipping a head is P(H) = 1/2. The probability of not flipping a head is P(no H) = 1 – P(H) = 1/2. The probability of not flipping a head in 3 flips (independent events) is

P(no H in 3 flips) = P(no H)^3 = (1/2)^3 = 1/8.

Therefore, the probability of flipping at least one head in 3 rolls is P(at least one H) = 1 – 1/8 = 7/8.

c. The probability of drawing a diamond is P(D) = 1/4. The probability of not drawing a diamond is P(no D) = 1 – P(D) = 3/4. The probability of not drawing a diamond in 5 draws (independent events) is

P(no D in 5 draws) = P(no D)^5 = (3/4)^5 = 0.24.

Therefore, the probability of drawing at least one diamond in 5 draws is P(at least one D) = 1 – 0.24 = 0.76.

d. The probability of drawing a ace is P(A) = 1/13. The probability of not drawing an ace is P(no A) = 1 – P(A) = 12/13. The probability of not drawing a ace in 10 draws (independent events) is

P(no A in 10 draws) = P(no A)^{10} = (12/13)^{10} = 0.45.

Therefore, the probability of drawing at least one ace in 10 draws is P(at least one A) = 1 – 0.45 = 0.55.

15. Better Bet for the Chevalier. We must use the "at least once" rule for this problem.

a. The probability of rolling a double six is 1/36. The probability of not rolling a double six is 35/36. The probability of not rolling a double six 25 rolls in a row is (35/36)^{25} = 0.4945. So the probability of rolling at least one double six in 25 rolls is 1 – 0.4945 = 0.5055. The odds are (slightly) better than even of winning.

b. If the Chevalier had played a slightly different game, he could have been expected to win in the long run and he may not have called Pascal to analyze his game.

17. Miami Hurricanes.

a. Given the historical record, the probability of Miami being hit by a hurricane in any given year is 1/40 = 0.025.

b. The probability of Miami being hit by a hurricane in two consecutive years is (1/40)^2 = 0.000625.

c. The probability of not being hit in a given year is 39/40. The probability of not being hit for ten consecutive years is (39/40)^{10} = 0.776. The probability of at least one hurricane in ten years is 1 – 0.776 = 0.224.

d. If the probability of Miami being hit by a hurricane were twice as high, or 1/20, then the probability of Miami being hit by a hurricane in two consecutive years is (1/20)^2 = 0.0025. The probability of not being hit in a given year would be 19/20. The probability of not being hit for ten consecutive years would be (19/20)^{10} = 0.599. So the probability of at least one hurricane in ten years would be 1 – 0.599 = 0.401, not quite twice the probability computed in part (c).

19. Poker Hands.

a. The number of different five-cards hands that can be dealt from a 52-card deck is $_{52}C_5 = 2,598,960$ (order does not matter).

b. The probability of being dealt one of 10, J, Q, K, or Ace of spades is 5/52. Having received one of those cards, the probability of getting one of the remaining four is 4/51 (there are 4 needed cards and 51 cards left in the deck). Having received two of the needed cards, the probability of getting one of the remaining three is 3/50 (there are 3 needed cards and 50 cards left in the deck). Continuing in this way, we see that the probability of being dealt a royal flush in spades (or any other specified suit) is

$$\frac{5}{52} \times \frac{4}{51} \times \frac{3}{50} \times \frac{2}{49} \times \frac{1}{48} = 3.8 \times 10^{-7}.$$

The odds of a royal flush in a specified suit are about 1 in 2.6 million (which is also 1 in the total number of card hands found in part (a)).

c. Since the four royal flushes are mutually exclusive, the probability of getting a royal flush in *any* suit is P(royal flush in spades) + P(royal flush in clubs) + P(royal flush in hearts) + P(royal flush in diamonds) = $4 \times (3.8 \times 10^{-7}) = 1.52 \times 10^{-6}$.

d. We found in part (a) that the total number of five-card hands is 2,598,960. The number of five-card hands with four aces is 48 (because the four aces can appear with any of the remaining 48 cards of the deck). So the probability of being dealt four aces is $48/(2,598,960) = 1.8 \times 10^{-5}$ or about one chance in 55,000. How many different five-card hands can have four of *any* kind (for example, four jacks or four 3s)? Just as there are 48 hands with four aces, there are 48 hands with any of the 13 different card denominations. So there are $13 \times 48 = 624$ different hands with four-of-a-kind. Therefore, the probability of being dealt any four-of-a-kind is $624/(2,598,960) = 2.4 \times 10^{-4}$ or about one chance in 4200.

21. Termite Genetics.
With a 90% chance of an offspring receiving the A gene from each parent, the probability of the AA pattern is $0.9 \times 0.9 = 0.81$. Similarly, the probability of the aa pattern is $0.1 \times 0.1 = 0.01$. Since the only other pattern is Aa, the probability of Aa must be $1 - 0.81 - 0.01 = 0.18$.

Unit 8D The Law of Averages

Overview

This unit is designed to strengthen your intuition about probabilities and to point out some common misconceptions about probabilities. It also deals with the important concept of expected value which is important to understand if you gamble or buy life insurance.

We start by discussing the **law of averages**. We know that the probability of tossing a head with a fair coin is 1/2. This does not mean that if you toss a coin twice that you will always get one head and one tail. It does not mean that if you toss a coin ten times that you will always get five heads and five tails. What we can say is that the more often you toss the coin, the closer you should expect the fraction of heads and tails to approach 50%. This is an example of the law of averages.

The law of averages is related to the outlook called the **gambler's fallacy**. Gamblers often feel that if they get behind or start losing money, then their luck will change and their chances will improve. The fact is that probabilities do not change just because someone is losing. Once behind, it is likely that you will stay behind or get even further behind.

The important concept in this chapter is **expected value**. Consider a situation in which there are several different outcomes. Each outcome has a known probability and a known cost or benefit. For example, a lottery may have a $10 prize, a $100 prize, a $1000 prize, and a grand prize of

$1,000,000. These are the four outcomes and each outcome has a certain probability. If you play this lottery many times, how much do you expect to win or lose in the long run? The answer is given by the expected value.

There is a general rule for computing expected values. In a situation with just two outcomes, the expected value is given by

$$\text{expected value} = \binom{\text{value of}}{\text{event 1}} \times \binom{\text{probability of}}{\text{event 1}} + \binom{\text{value of}}{\text{event 2}} \times \binom{\text{probability of}}{\text{event 2}}$$

The remainder of the unit is devoted to applications of expected value as it arises in life insurance policies, lotteries, and the so-called house edge in casino gambling. There are some important practical lessons to be earned here.

Key Words and Phrases

law of averages   gambler's fallacy   expected value
house edge

Key Concepts and Skills

- explain the law of averages.
- explain the gambler's fallacy.
- compute the expected value in situations with two or more outcomes.
- understand the house edge and compute it in specific situations.

Solutions Unit 8D

1. Roulette.

a. There are 18 black numbers and a total of 38 numbers on the roulette wheel. The probability of a black (or a red) number on a single spin is 18/38 = 9/19 = 47%.

b. Although the probability of getting black on any spin is 47%, there is no way to predict the actual outcome on any single spin. Similarly, with only three spins, you cannot make any reasonable prediction about how many of the spins will come up red.

c. The law of averages tells us that as the game is played more and more times, the percentage of times that the wheel comes up black should get closer to the 47% probability. Thus, in 100,000 tries, the wheel should come up red close to 47% of the time, or about 47,000 times.

3. Dice Rolling.

a. The probability of rolling a 6 in one roll of a fair die is 1/6 = 16.7%.

b. Although the probability of getting a 6 on any roll is 1/6, there is no way to predict the actual outcome on any single roll. With only 12 rolls, you might predict that less than 3 or 4 rolls will show a 6.

c. The law of averages tells us that if a die is rolled more and more times, the percentage of times that the die comes up 6 should get closer to the 16.7% *a priori* probability. Thus, in 1000 tries, the die should show 6 close to 16.7% of the time, or about 160 to 170 times.

5. Gambler's Fallacy with Coin Tossing.

a. Heads has come up 45% of the time in the first 100 tosses. You have won $45 and paid out $55 for a net loss of $10.

b. One would expect with more tosses that the percentage of heads would get closer to 50%; and in this situation it does. If 47% of the first 300 tosses are heads, there have been 141 heads and 159 tails. Thus your net gain is $141 − $159 = −$18 — an even greater loss!

c. If you were to break even after 400 tosses, you would have to get 200 heads. With 141 heads in the first 300 tosses, this means you would have to get another 59 heads in the next 100 tosses. This is rather unlikely to occur because the probability of heads is 50%, significantly less than 59%.

d. The above situation illustrates why if you lose and continue to play, you will not recover your losses even though the percentage of wins increases. This is the crux of the gambler's fallacy.

7. Behind in Coin Tossing: Can You Catch Up?

a. On the next toss you could get a head which makes a total of 39 heads and 62 tails (difference 23). Equally likely, you could get a tail which makes a total of 38 heads and 63 tails (difference 25). So the difference between the number of heads and tails could increase or decrease by one with equal probability.

b. On each successive toss, the difference is equally likely to increase as decrease. Therefore, after 1000 tosses, the original difference of 24 (which represents the gambler's losses) is just as likely to increase as it is to decrease.

c. Once you have fewer heads than tails (say 38 heads vs. 62 tails), the difference between the number of heads and tails is equally likely to increase as decrease, so the deficit of heads is likely to remain.

d. As argued in part (c), once you have fewer heads than tails, you are likely to maintain fewer heads than tails in future tosses. So if heads represent winners, you are more likely remain on the losing side. The gambler's fallacy would advise you to continue even though you are losing. In reality, you are more likely to remain on the losing side if you continue.

9. Should You Play? The probability of getting a double number when you roll two dice is 6/36 =1/6. The probability of not getting a double number is 1 − 1/6 = 5/6. The value of getting a double number is $8 and the value of not getting a double number is −$1. The expected value for this game is

$$\frac{1}{6} \times \$8 + \frac{5}{6} \times (-\$1) = \$0.50.$$

On average you could expect to win 50 cents per game. Yes, you should play! However you should not expect to see this gain if you play only once. After a large number of games, your winnings should approach the expected value.

11. Expected Value and Insurance. We begin by making a table of all the "events" associated with this insurance policy. Note that revenues for the company are positive and costs are negative.

event	value	probability
sale of policy for $200	$200	1
claim of $2000	−$2000	1/50
claim of $1000	−$1000	1/20
claim of $500	−$500	1/10

There are four events in this situation, so we find the expected value by adding the four products *value × probability*:

$$(\$200 \times 1) - (\$2000 \times \frac{1}{50}) - (\$1000 \times \frac{1}{20})$$

$$- (\$500 \times \frac{1}{10}) = \$60.$$

On average, the company should expect to earn $60 per policy. However, the company should expect the *actual* earnings to be close to this average only if it sells many thousands of policies.

13. Lottery Expectations. To compute the expected value we add the products *value × probability* for each outcome. Notice that one outcome is buying the $1 ticket which has probability equal to 1. The expected value of the lottery ticket is

$$-(\$1 \times 1) + (\$2 \times \frac{1}{20}) + (\$5 \times \frac{1}{100}) + (\$10 \times \frac{1}{500}) +$$

$$(\$1000 \times \frac{1}{10,000}) + (\$10,000 \times \frac{1}{100,000}) + (\$10^6 \times \frac{1}{10^7}) = -\$0.53.$$

You can expect to lose about 53 cents per ticket, at least in the long run.

15. Dice Rolling Expectations. Each outcome has the same probability, namely 1/6. To compute the expected value we add the products *value × probability* for each outcome. The expected value of the game is

$$\left(\$1 \times \frac{1}{6}\right) + \left((-\$2) \times \frac{1}{6}\right) + \left(\$3 \times \frac{1}{6}\right) + \left((-\$4) \times \frac{1}{6}\right)$$

$$+ \left(\$5 \times \frac{1}{6}\right) + \left((-\$6) \times \frac{1}{6}\right) = -\$0.50.$$

In the long run can you expect to lose about 50 cents per game.

17. House Edge in Roulette. The probability of winning with three numbers is

P(4 or 8 or 16 or 32) = P(4) + P(8) + P(16) + P(32) = 1/38 + 1/38 + 1/38 + 1/38 = 4/38.

If you win, the casino pays you $35, but you lose $3 on your three losing bets, so you net winnings are $32. The probability of losing with three numbers is 34/38, in which case you lose $4. The expected earnings for the game are

$$\left((-\$4) \times \frac{34}{38}\right) + \left(\$32 \times \frac{4}{38}\right) = -\$0.211.$$

On average you will lose 21.1 cents on a $4 bet. So the house edge (the amount won per dollar bet) is $0.211/4 = $0.053 which is the same house edge as for a bet on a single number.

19. House Edge in Blackjack.

a. The probability of losing at blackjack is 0.507 and the probability of winning at blackjack is $1 - 0.507 = 0.493$. Suppose you bet $1 on a single game. Then the value of winning is $1 and the value of losing is –$1. So the expected value to you is

$$(-\$1) \times 0.507 + \$1 \times 0.493 = -\$0.014.$$

You can expect to lose 1.4 cents for every dollar you bet. The house edge is 1.4 cents per dollar.

b. If you play 100 games with a $1 on each game, you will lose $100 × 1.4 cents/dollar = $1.40.

c. If you play 100 games with a $5 on each game, you will lose $500 × 1.4 cents/dollar = $7.

d. With $1,000,000 in bets in an evening, the expected earnings of the casino are $1,000,000 × 1.4 cents/dollar = $14,000.

21. Profitable Casino. The house edge is the amount the casino earns per dollar bet. If $1 billion is bet in a year, the casino can expect to earn $0.085 per dollar × $1 billion = $85 million.

Unit 8E Probability, Risk, and Coincidence

Overview

This concluding unit of the chapter is an amusing collection of topics, all related in some way to probability. The first topic is estimating risk. You often hear statements such as "you are more likely to get struck by lightning than die in an airplane crash." How are such statements concocted? And to what extent are they true? We will explore how such comparisons of risk are made, particularly as they apply to **vital statistics** — data about births and deaths.

You will see that these problems are examples of empirical probability. If we know the number of deaths due to, say, heart disease, we can estimate the chances of a person dying from heart disease. If we know the number of deaths due to automobile accidents, we can estimate the chances of a person dying in an automobile accident. With these two facts, it is possible to compare the risks of automobiles and heart disease. Clearly, risk analysis and decision making are closely related.

Related to vital statistics are the subjects or **mortality** and **life expectancy**. We will look at graphs of death rate and life expectancy as they change with age. The interpretation of these graphs reveals a few surprises. Why should life expectancy actually increase with age? Does the death rate increase steadily with age? Needless to say, these graphs have many implications.

The second topic of the unit is **coincidence**. Should you really be surprised when you win a raffle drawing at a basketball game? Should you really be surprised that after an evening of playing cards you are dealt a 13-card hand with eight hearts? Many so-called coincidences are bound to happen to someone, but we are always surprised when they happen to us.

We discuss the famous birthday problem and show that if there are 25 people in a room, then there is better than a 50% chance that two people have the same birthday. We also explore the phenomenon of **streaks**, particularly in sports, when a player repeats a certain success for many consecutive games. Should you really be surprised?

Key Words and Phrases

risk analysis	vital statistics	mortality
life expectancy	coincidence	

Key Concepts and Skills

- determine the probability of an event given the number of people who experience the event.
- compare the risks of two events or two causes of death.
- interpret life expectancy and mortality tables.
- understand why some coincidences should not be surprising.

Solutions Unit 8E

1. Automobile Risk.

a. If we extend the graph of Figure 8-9 it suggests that there were about 45,000 automobile deaths in 1998. Using 270 million as the U.S. population for 1998, the probability of dying in an automobile accident was:

$$1998: \quad \frac{45,000}{270 \text{ million}} = 1.66 \times 10^{-4}.$$

b. We have to solve an "at least once" problem! From part (a), the probability of dying in a single year is 1.66×10^{-4}. The probability of not dying in a single year is $1 - (1.66 \times 10^{-4}) = 0.99983$. The risk of not dying over a ten-year period is $(0.99983)^{10} = 0.99834$. Therefore the probability of dying (at least once) in a ten-year period is $1 - 0.99834 = 0.00166$ or about 17 in 10,000. This calculation assumes the risk remains constant for 10 years.

c. As in part (b), we must solve an "at least once" problem. From part (a), the probability of dying in a single year is 1.66×10^{-4}. The probability of not dying in a single year is $1 - (1.66 \times 10^{-4}) = 0.99983$. The risk of not dying over a 50-year period is $(0.99983)^{50} = 0.99153$. Therefore the probability of dying (at least once) in a 50-year period is $1 - 0.99153 = 0.00847$ or

about 8 in 1000. This calculation assumes the risk remains constant for 50 years.

3. Non-Commercial Aviation Safety.
The fatality rate was 631 fatalities/ 23.7 million flight hours = 2.66×10^{-5} fatalities per flight hour or 2.66 fatalities per 100,000 flight hours. This rate is almost the same as the fatality rate for commercial flights (2.77×10^{-5} fatalities per flight hour).

5. High/Low U.S. Birth Rates.

a. In Utah there were 39,577 births per year or

$$39{,}577\frac{\text{births}}{\text{year}} \times \frac{1 \text{ year}}{365 \text{ days}} = 108\frac{\text{births}}{\text{day}}.$$

In Maine there were 13,896 births per year or

$$13{,}896\frac{\text{births}}{\text{year}} \times \frac{1 \text{ year}}{365 \text{ days}} = 38\frac{\text{births}}{\text{day}}.$$

b. The Utah birth rate for the year was 20 births per 1000 people and there were 39,577 births. One way to find the total population is by setting up a proportion:

$$\frac{20 \text{ births}}{1000 \text{ people}} = \frac{39{,}577 \text{ births}}{\text{total population}}.$$

Solving this equation for total population, we see that

$$\text{total population} = \frac{39{,}577 \text{ births} \times 1000 \text{ people}}{20 \text{ births}} = 1{,}978{,}850.$$

The 1995 population of Utah was almost 2 million people.

The Maine birth rate for the year was 11.2 births per 1000 people and there were 13,896 births. Let's use units this time to find the total population Maine. We have

$$13{,}896 \text{ births} \times \frac{1000 \text{ people}}{11.2 \text{ births}} = 1{,}240{,}714 \text{ people.}$$

The 1995 population of Maine was about 1.2 million people.

7. U.S. Birth and Death Rates.

a. The number of births in the United States in 1997 was

$$\frac{14.7 \text{ births}}{1000 \text{ people}} \times 270 \text{ million people} = 4.0 \text{ million births.}$$

b. The number of deaths in the United States in 1997 was

$$\frac{8.7 \text{ deaths}}{1000 \text{ people}} \times 270 \text{ million people} = 2.3 \text{ million deaths.}$$

c. The net population gain was the number of births minus the number of deaths or 4.0 million − 2.3 million = 1.7 million people in one year (excluding immigration).

d. The difference between the population increase computed in part (c) and the actual population increase is due to immigration. Therefore the population increase due to immigration was 2.4 million − 1.7 million = 0.7 million. This figure is the difference between the actual population increase and the population increase due to births. The fraction of the population change due to immigration was (0.7 million)/(2.4 million) = 0.292 = 29.2%. About thirty percent of the population increase was due to immigration.

9. Interpreting the Vital Statistics Table.

a. The empirical probability of death by any particular cause is the number of deaths attributed to that cause divided by the total population of 260 million.

$$P(\text{death by homicide}) = \frac{20{,}000}{260{,}000{,}000} = 7.7 \times 10^{-5}$$

$$P(\text{death by auto}) = \frac{44{,}000}{260{,}000{,}000} = 1.7 \times 10^{-4}.$$

About 8 people per 100,000 died by homicide. In contrast, 17 people per 100,000 died in automobile accidents. Thus, overall, the probability of death by automobile accidents is

$$\frac{1.7 \times 10^{-4}}{7.7 \times 10^{-5}} = 2.2$$

times greater than the probability of death by homicide.

b. From part (a), the death rate due to homicide is 7.7×10^{-5} deaths per person. To find deaths per 100,000

people, we *multiply* by $100{,}000 = 10^5$. The death rate is 7.7 deaths per 100,000 people.

c. If we assume the death by homicide and death by automobile accident are mutually exclusive, then the probability of death by homicide or automobile is

$$P(\text{death by homicide}) + P(\text{death by automobile})$$
$$= 7.7 \times 10^{-5} + 1.7 \times 10^{-4} = 2.5 \times 10^{-4},$$

or 0.025% or 2.5 chances in 10,000.

11. Birthday Coincidences.

a. Following Example 6 of the text, we begin by recognizing that, with 365 days in a year, the probability that any particular student has *your* birthday is 1/365; the probability that a particular student does *not* have your birthday is 364/365. Thus the probability that *none* of the 14 other students in the class has your birthday is

$$\left(\frac{364}{365}\right)^{14} = 0.96.$$

The chance that at least one person in the class *does* have your birthday is

$$1 - 0.96 = 0.04,$$

or about 4%, or about 1 in 25.

b. Following Example 7 of the text, we begin by considering just two students from the class. The first student has a birthday on 1 of the 365 days in a year. Thus the probability that the second student has a different birthday is

$$\frac{\text{number of possible } \textit{different} \text{ birthdays}}{\text{total number of possible birthdays}} = \frac{364}{365}.$$

Now consider a third student. The probability that all 3 students have different birthdays is a *dependent* probability: the probability that the third student has a different birthday *given that* the first two students have different birthdays.

$$P(3 \text{ different birthdays})$$
$$= P\left(\begin{array}{c}\text{third student has}\\\text{different birthday}\end{array}\right) \times P\left(\begin{array}{c}\text{first 2 students have}\\\text{different birthdays}\end{array}\right)$$

If the first two students have different birthdays, 2 of the 365 days in the year are "taken," leaving a probability of 363/365 that the third student has a different birthday. Thus the probability that all 3 have different birthdays is

$$P(3 \text{ different birthdays}) = \frac{363}{365} \times \frac{364}{365}.$$

Similarly, if the first three students all have different birthdays, the probability that a fourth student has a different birthday from any of the first three is 362/365. Continuing in this way, the probability that all 15 students have different birthdays is

$$\frac{364}{365} \times \frac{363}{365} \times \frac{362}{365} \times \frac{361}{365} \times \frac{360}{365} \times \frac{359}{365} \times \frac{358}{365} \times \frac{357}{365}$$

$$\times \frac{356}{365} \times \frac{355}{365} \times \frac{354}{365} \times \frac{353}{365} \times \frac{352}{365} \times \frac{351}{365} = 0.747.$$

There is a 0.75 chance that no two students in the class of 15 have the same birthday. The chance that at least two students *do* have the same birthday is $1 - 0.75 = 0.25$.

In other words, there's a 25% chance that at least two of the 15 students have the same birthday! Note that this is a much higher probability than the 4% chance that someone in the class has *your* birthday (see part (a)).

13. Hot Streaks.

a. The probability of one person winning five games in a row is $(0.48)^5 = 0.025$.

b. The probability of one person winning ten games in a row is $(0.48)^{10} = 0.00065$.

c. If 2000 people were playing the game then (using the result of part (a)), $0.025 \times 2000 = 50$ people could be expected to win five games in a row.

d. If 2000 people were playing the game then (using the result of part (b)), $0.00065 \times 2000 = 1.3$ people could be expected to win ten games in a row; about 1 person of the 2000 could be expected to have a hot streak.

17. Joe DiMaggio's Record.

a. With a batting average of .400, the probability of not getting a hit in one at-bat is 0.6. The probability of not getting a hit in four at-bats is $(0.6)^4 = 0.1296$. The probability of getting at least one hit in four at-bats is $1 - 0.1296 = 0.8704$.

b. Using the result of part (b), the probability of getting at least one hit in each of 56 consecutive games is $(0.8704)^{56} = 0.0004$ or 4 in 10,000 or 1 in 2500.

c. If we change the batting average to .300, then the probability of having a 56-game hitting streak is (combining the steps of parts (a) and (b))

$$((1 - (0.7)^4))^{56} = 2 \times 10^{-7}$$

or 2 in ten million.

d. If we assume that there are 50 good hitters (batting average better than .300) each year and baseball has been played for 100 years, then there have been 5000 good hitters, which is not enough hitters to expect a 56-game hitting streak. If we could argue that there are 25 hitters each season with averages above .400, there would be 2500 good hitters over the 100-year history. Then we could expect one hitter to have had a 56-game streak.

9. PUTTING STATISTICS TO WORK

Overview

We now return to the subject of statistics, having already introduced some of the qualitative aspects in Chapter 2. The treatment of statistics in this chapter is more quantitative and involves some numerical calculations. In Unit 9A we present several methods for characterizing data sets, including the calculation of the mean, median, mode, and standard deviation. Unit B covers the topic of linear regression, a tool that allows us to identify possible correlations between two variables. In Unit 9C, we introduce the normal distribution which lies at the heart of many statistical methods. Finally, Unit 9D presents several statistical case studies that illustrate statistical issues in today's world.

Unit 9A Characterizing Data

Overview

In Chapter 2 we discussed the processes used to collect data. In this unit we explore the methods used to describe and analyze data. Typical data sets often contain hundreds or thousands of numbers, so one goal is to summarize the data set in compact and meaningful ways.

We begin by considering frequency data sets that give the number of people or objects in the sample that fall into the same category. For instance, you might sample your class to find the number of students with various hair colors or your teacher might have a grade distribution that shows how many students have various grades. Frequency data can be summarized in a **frequency table** that shows the **frequency**, **relative frequency**, and **cumulative frequency** of each data value. Frequency data can also be displayed graphically using a **dotplot** or a **histogram**. If a data set has many data categories or a continuous range of categories, then it is necessary to group or **bin** the data before making a table or a histogram.

These techniques condense that data to some extent. But for a very concise summary of the data, it is common to compute the **mean**, **median**, and/or **mode** of the data. These one-number summaries are defined as follows:

- The *mean* of a data set is calculated by the formula

$$\text{mean} = \frac{\text{sum of all values}}{\text{total number of values}}.$$

- The *median* is the middle score in the data set. Note that there will be two "middle" values if a data set has an even number of data points; if the two middle values are different, the median lies halfway between them.
- The *mode* is the most common score in a data set. A data set may have more than one mode, or no mode.

Having determined where the "center" of the data set lies, it is next useful to describe the shape of the data: are the data values symmetric? are they weighted to the right or left? These questions can be answered qualitatively by finding whether the data values are **positively** or **negatively skewed**.

All of the above ideas can be captured in one compact description of the data called the **five-number summary,** which consists of the **median,** the **upper** and **lower quartiles,** and the **highest** and **lowest data value.** The five-number summary can be displayed nicely with a **box plot.**

More insight into a data set can be gained when we quantify the spread or **dispersion** of the data. The key quantities for this purpose are the **variance** and the **standard deviation.** You will see detailed calculations of these quantities.

All of these methods and measures are accompanied by numerical examples in the text. Be sure you understand how these calculations are done and then practice on the data sets in the problems. It is also helpful to learn your calculator's statistical capabilities. Many calculators have these statistical functions built in to them. Needless to say, these functions can save you a lot of work and let you concentrate on the meaning of the numbers.

Key Words and Phrases

exploratory data analysis	inferential statistics	raw data
frequency	frequency table	relative frequency
cumulative frequency	dotplot	binning data
mean	median	mode
single-peaked	bimodal	symmetric
outliers	positively skewed	negatively skewed
five-number summary	upper quartile	lower quartile
dispersion	box plot	deviation
variance	standard deviation	

Key Concepts and Skills
- explain the purpose of exploratory data analysis and inferential statistics.
- given a frequency data set, construct a frequency table showing relative and cumulative frequency.
- bin a data set.
- compute the mean, median, mode, five-number summary, box plot, and standard deviation of a frequency data set.
- determine the qualitative shape of a data set.

Solutions Unit 9A

1. Term Paper Results. A frequency table for the $N = 22$ scores after binning is shown below, as are the pie chart and bar graph.

Grade	Frequency	Relative Frequency	Cumulative Frequency
A	3	3/22 = 13.6%	3
B	7	7/22 = 31.8%	10
C	7	7/22 = 31.8%	17
D	2	2/22 = 9.1%	19
F	3	3/22 = 13.6%	22
Total	**22**	**99.9%**	**22**

b., c. The pie chart and the bar graph for the data are shown below.

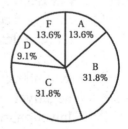

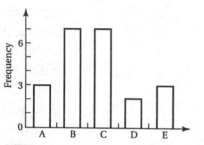

3. Exploring Tables. Here is the completed table. The fact that the total of the cumulative frequencies is 10 says that there is a total of 10 times. Also the relative frequencies must add up to 1.

Time	Frequency	Relative Frequency	Cumulative Frequency
10.0	1	0.1	1
10.1	3	0.3	4
10.2	2	0.2	6
10.3	4	0.4	10
Total	**10**	**1.0**	**10**

5. Mean, Median, and Mode.

a. The mean of the 10 scores is

$(5 + 5 + 6 + 7 + 7 + 9 + 9 + 9 + 9 + 10)/10 = 7.6$.

b. The median (or middle score) is 8 (halfway between 7 and 9).

c. The mode (most common score) is 9.

7. Commuter Statistics.

a. The mean of the 10 times is 25.16 minutes.

b. The median is between 25.3 minutes and 26.3 minutes, or 25.8 minutes.

c. The five-number summary of the scores is

    the median is between 25.3 minutes and 26.3
    minutes (or 25.8 minutes)

    the upper quartile is 26.5 minutes

    the lower quartile is 23.8 minutes

    the fastest time is 22.1 minutes

    the slowest time is 27.0 minutes.

d. The box plot is shown below.

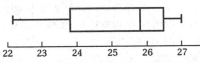

e. The table below shows the calculation of the standard deviation. The mean from part (a) is 10.92

Time (sec)	Deviation	Squared deviation
22.1	22.1 – 25.16 = –3.06	9.36
23.4	23.4 – 25.16 = –1.76	3.10
23.8	23.8 – 25.16 = –1.36	1.85
24.1	24.1 – 25.16 = –1.06	1.12
25.3	25.3 – 25.16 = 0.14	0.0196
26.3	26.3 – 25.16 = 1.14	1.30
26.3	26.3 – 25.16 = 1.14	1.30
26.5	26.5 – 25.16 = 1.34	1.80
26.8	26.8 – 25.16 = 1.64	2.69
27.0	27.0 – 25.16 = 1.84	3.39
Total	0.0	25.93

The variance is found by dividing the sum of the squared deviations by the number of times *minus 1* (or 9). So the variance is 25.93/9 = 2.88. The standard deviation is the square root of the variance or $\sqrt{2.88}$ = 1.70.

f. The standard deviation makes sense. Notice that an interval one standard deviation on either side of the mean extends from 23.46 minutes to 26.86 minutes which includes 7 of the 10 times.

9. Deviations in Height.

a. The five-number summary of the scores is:

    the median is 172 cm

    the upper quartile is 188 cm

    the lower quartile is 166 cm

    the largest height is 190 cm

    the smallest height is 145 cm.

b. The box plot is shown below.

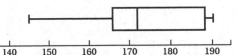

c. The mean of the seven heights is 172.6 cm.

d. The table below shows the calculation of the standard deviation. The mean from part (c) is 172.6 cm.

Height (cm)	Deviation	Squared deviation
145	145 – 172.6 = –27.6	762
166	166 – 172.6 = –6.6	43.6
169	169 – 172.6 = –3.6	13.0
172	172 – 172.6 = –.6	0.36
178	178 – 172.6 = 5.4	29.2
188	188 – 172.6 = 15.4	237
190	190 – 172.6 = 17.4	303
Total		1388

The variance is found by dividing the sum of the squared deviations by the number of heights *minus 1* (or 6) which is 1388/6 = 231. The standard deviation is the square root of the variance or $\sqrt{231}$ –15.2.

e. An interval one standard deviation on either side of the mean extends from 157 cm to 188 cm which includes all but two of the seven heights. So the standard deviation makes sense for this small distribution.

11. Air Force Cadets.

a. The data table with the relative and cumulative frequencies is shown below.

Reflex Time	Frequency	Rel. Freq.	Cum Freq.
1.1	5	0.1	5
1.2	10	0.2	15
1.3	20	0.4	35
1.4	8	0.16	43
1.5	4	0.08	47
1.6	2	0.04	49
1.7	1	0.02	50
Total	50	1.0	50

b, c, d. The histogram, frequency line chart, and cumulative frequency line chart are shown below.

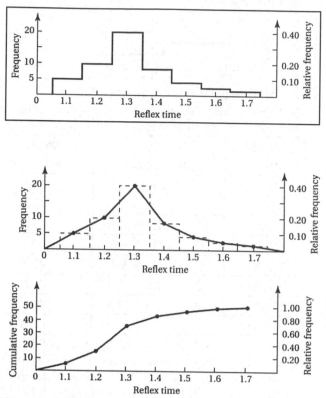

e. The mean is 1.312 seconds.

f. The median of the data is 1.3 seconds.

g. The mode of the data is 1.3 seconds.

h. The variance is 0.017. The standard deviation is 0.13.

i. Most of the data are at the low end of the distribution, while a few outliers at the high end push the mean above the median and mode. This distribution is positively skewed.

P 13. Income and Degrees.

a. These data were probably collected by a survey of a randomly selected sample. Perhaps the most significant sampling error was misreporting of data: some people tend to overestimate their income.

b. Income data is usually positively skewed because there is a lower limit to income, but virtually no upper limit. The mean would be higher than the median because it is pulled upward by high outliers.

15. Skewness. A large number of high scores will result in a high median and mode, while the few lower scores will give the distribution a low mean. Therefore, the distribution is skewed negatively.

17. Skewness. Most cars tend to go uniformly fast, while a few cars go slower. This pattern will produce a negatively skewed distribution of speeds. However one might argue that as many cars go excessively fast as excessively slow, which would produce a symmetric distribution.

19. Skewness. There aren't any factors (are there?) that would cause the distribution of pizza eaters to be skewed in either direction. The outliers (heavy and light eaters) would be distributed evenly on either side of the mean. The distribution is symmetric.

The Shape of Data.

21.

For data set (i):

a. The five-number summary is:

    the median is 75 minutes

    the upper quartile is 80 minutes

    the lower quartile is 60 minutes

    the largest time is 110 minutes

    the lowest time is 45 minutes.

b. The data set is bimodal because it has two distinct peaks.

For data set (ii):

a. The five-number summary is:

    the median is 2 months

    the upper quartile is 3 months

    the lower quartile is 1 months

    the largest time is 11 months

    the lowest time is 1 month.

b. The data set is positively skewed because there is a lower limit to the variable failure time (0 months), but no upper limit.

For data set (iii).

a. The five-number summary is:

    the median is 90 kilograms

    the upper quartile is 100 kilograms

    the lower quartile is 85 kilograms

    the largest weight is 125 kilograms

    the lowest weight is 60 kilograms.

b. The data set is symmetric.

Unit 9B Linear Regression Models

Overview

This short unit has one important topic: determining whether there is a correlation between two variables. Suppose you collect data on the number of hours studied per week and the grade received for each student in your class. Then you could make a **scatter plot** by graphing the variables *hours studied* and *grade received* for each student. If the points on the scatter plot fall close to a line, you have reason to believe that the variables are well correlated. In this case, you might observe that an increase in study time generally leads to an increase in grades. The process just described is called **linear regression**.

The linear regression that we will do in this unit is visual: we will draw the line that *appears* to fit a set of data most closely. In more advanced statistics courses, you could study a specific method for finding the line that gives the best fit to the data.

The measure of how well two variables are correlated is called the **correlation coefficient**. Denoted r, the correlation coefficient lies between -1 and 1.

- If $r = 1$, the two variables are *perfectly* and *positively* correlated. All the data points lie precisely on the best-fit line, and an increase in one variable means an increase in the other.
- If r is between 0 and 1, the two variables are *positively* correlated, but not perfectly. That is, the data points do not all lie precisely on the best-fit line, but an increase in one variable *tends* to mean an increase in the other. The stronger this tendency, the closer the correlation coefficient is to 1.
- If $r = 0$, the two variables are completely *uncorrelated*: there is no linear relationship between them.
- If r is between 0 and -1, the two variables are *negatively* correlated, but not perfectly. That is, an increase in one variable *tends* to mean a *decrease* in the other. The stronger this tendency, the closer the correlation coefficient is to -1.
- If $r = -1$, the two variables are *perfectly* and *negatively* correlated. All the data points lie precisely on the best-fit line, and an increase in one variable means a *decrease* in the other.

An **outlier** is a data point that does not fit the general pattern of a data set or lies far from the regression line. A single outlier can affect the regression line and the correlation coefficient significantly. The unit closes with some examples of visual linear regression and a remark on **nonlinear regression** in which data sets are fit with curves other than straight lines.

Key Words and Phrases

scatter plot   linear regression   correlation coefficient

outlier   nonlinear
                      regression

Key Concepts and Skills

- explain the general goal of linear regression.
- draw a scatter plot for a two-variable data set.
- construct the regression line for a data set visually.
- understand the meaning of the correlation coefficient and estimate it for a given scatter plot.
- understand the effect of outliers.

Solutions Unit 9B

5. Television Time.

a. The data were collected through extensive surveys of TV audiences. Care must be used when comparing surveys over many years (in this case, over 40 years). Sampling and polling techniques have changed and the comparisons between years may not be reliable. The apparent increase in TV viewing hours may be attributable, in part, to differences in survey techniques.

b. The histogram and line chart for the data are shown below.

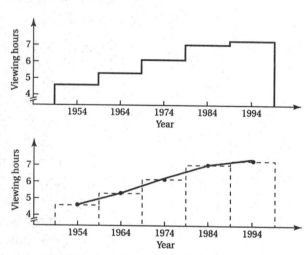

c. A regression line (not shown) would fit the first four data points very well. If the 1994 figure is included the fit is not as good.

d. Any projection beyond 1994 is not reliable since the 1994 data point suggests that there is a significant slowing in the increase of the TV viewing. The trend may not continue to be linear.

7. Diving Reflex.

a, b. The scatter plot for the data is shown below. A visual best-fit line is also drawn on the plot.

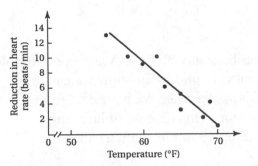

c. There is a strong negative correlation between the variables and the regression line fits the data very well.

d. A good estimate for the correlation coefficient is $r = -0.90$. It suggests that the diving reflex is a real effect.

9. Women in the Labor Force.

a. Three separate time-series graphs could be made to display these labor data. However, since two of the variables use the same units (%), it is possible to put all three time series on a single graph, provided the scales are read carefully.

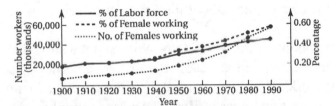

b. Although regression lines have not been drawn on the graph, it appears that all three curves have a curving upward trend (the rate of change is itself increasing). This trend usually means that straight line fits will not give an accurate representation of the data. They would most likely not give a good prediction of the variables in 2010.

c. Looking at 1995 data, we see that about 59% of the female population work and that 46% of the work force consists of women. We can also assume that half of the entire population consists of women. From this information can we say that more than 59% of men work. But can we find the exact breakdown of the

entire population among men/women and workers/non-workers?

d. We use the facts that about 59% of the female population work and that 46% of the work force consists of women. We can also assume that half of the entire population consists of women. The first observation is that $0.59 \times 0.50 = 0.295$ or 29.5% of all people are women workers (in other words, 59% of the 50% of all people who are women are also workers). This implies that 50% − 29.5% = 20.5% of all people are women non-workers.

Now what about the men's side of the picture? The 29.5% of the population consisting of women workers represents 46% of all workers. Therefore, 0.295/0.46 = 64.1% of the population are workers. Since 29.5% of the population consists of women workers, 64.1% − 29.5% = 34.6% of the population consists of men workers. This leaves 50% − 34.6% = 15.4% of the population as men non-workers.

d. We can make a pie chart showing this breakdown.

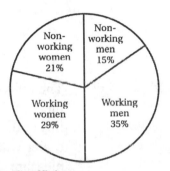

Correlation Coefficients.

11. The data points in this scatter plot are scattered in a cloud, meaning they are essentially uncorrelated. A reasonable correlation coefficient would be close to zero, say, $r = 0.05$, or $r = -0.05$, $r = 0.00$.

13. The data points in this scatter plot are loosely grouped along a rising line. The data are weakly correlated and the correlation is positive. A reasonable correlation coefficient would be positive and near 0.5, say, $r = 0.50$.

15. The data points in this scatter plot are scattered randomly, meaning they are essentially uncorrelated. A reasonable correlation coefficient would be $r = 0.00$.

Unit 9C Statistical Inference

Overview

This unit is about the normal distribution, which plays an important role in all of statistics. The goal of this unit is to understand how we can draw some conclusions about an entire population when we have data from a relatively small sample of that population. This process of inferring something about the population based on a sample is often called **inferential statistics**.

We begin by introducing the **normal distribution**, a whole family of symmetric frequency distributions that are characterized by their mean and their standard deviation. Many variables, such as test scores, heights, weights, and other physical characteristics, follow a normal distribution closely. One of the most important results about the normal distribution is the **68-95-99.7 Rule:**

- About 68% (actually 68.3%) of the data points fall within 1 standard deviation of the mean.
- About 95% (actually 95.4%) of the data points fall within 2 standard deviations of the mean.
- About 99.7% of the data points fall within 3 standard deviations of the mean.

Thus if we know that a variable is distributed normally and we know its mean and standard deviation, we can say a lot about where a particular score lies in the distribution. Working with the normal distribution is made much easier with the use of *z*-scores and **percentiles**. The *z*-score simply measures how many standard deviations a particular data value is above or below the mean. Once the *z*-score is known, the percentile can be found using Table 9.7 (which appears in all statistics books). Similarly, given a percentile, the corresponding *z*-score can be found.

The unit next returns to **margins of error** and **confidence intervals**, which were introduced in Unit 2A. The problem of interest is a very common one for opinion polls and surveys. Suppose you want to know the fraction (or percentage) of people in a population that hold a particular opinion or have some trait. The usual practice is to select a sample and determine what fraction (or percentage) of people in the sample hold a particular opinion or have that trait; this is called the **sample statistic**. Knowing the sample statistic, what can you conclude about the entire population?

What you will see in the unit (with plenty of explanation and examples) is that if the sample has n individuals, then the fraction of interest for the population lies within $1/\sqrt{n}$ of the sample statistic – with 95% certainty. The quantity $1/\sqrt{n}$ is called the **margin of error** and the interval that extends $1/\sqrt{n}$ on either side of the sample statistic is called the **95% confidence interval**. For example, if in a sample of $n = 10,000$ people, 45% have blond hair, then the margin of error is $1/\sqrt{10,000} = 0.01 = 1\%$ and the margin of error is 45% − 1% to 45% + 1% or 44% to 46%. We can conclude with 95% certainty that between 44% and 46% of the entire population has blond hair. This is one of the most basic results of inferential statistics. It allows us to say something about an entire population based on what we learn from a sample.

Key Words and Phrases

normal distribution	68-95-99.7 rule	percentile
z-score	margin of error	confidence interval

Key Concepts and Skills
- understand the goal of inferential statistics.
- draw a normal distribution curve with a given mean and standard deviation.
- use the 68-95-99.7 Rule to analyze a normal distribution.
- convert z-scores to percentiles and vice versa.
- construct a margin of error and confidence interval for a given sample statistic with a known sample size.

Solution Unit 9C

1. Normal Distributions. Recall that by the 68-95-99.7 Rule, about 68% (actually 68.3%) of the data points fall within 1 standard deviation of the mean, about 95% (actually 95.4%) of the data points fall within 2 standard deviations of the mean, and about 99.7% of the data points fall within 3 standard deviations of the mean.

a. The range from 50 to 70 covers one standard deviation on either side of the mean. Therefore 68.3% of the scores lie in this range.

b. The range from 60 to 70 covers one standard deviation above the mean. Therefore half of 68.3% of the scores or about 34% of the scores lie in this range.

c. The range from 50 to 60 covers one standard deviation below the mean. Therefore half of 68.3% of the scores or about 34% of the scores lie in this range.

d. The range from 40 to 80 covers two standard deviations on either side of the mean. Therefore 95% of the scores lie in this range.

e. The range from 30 to 90 covers three standard deviations on either side of the mean. Therefore 99.7% of the scores lie in this range.

f. The range from 30 to 60 covers three standard deviations below the mean. Therefore half of 99.7% of the scores or about 49.85% of the scores lie in this range.

3. Graphing the Normal Distribution.

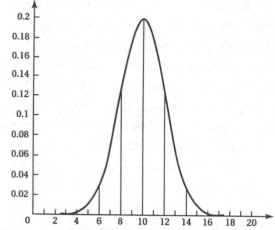

a. By the 68-95-99.7 Rule, about 68% (actually 68.3%) of the data points fall within 1 standard deviation of the mean. Thus 68% of the observations lie between 8 and 12. These observations lie between the vertical lines at 8 and 12 on the graph.

b. By the 68-95-99.7 Rule, 95% of the data points fall within 2 standard deviations of the mean. Thus 95% of the observations lie between 6 and 14. These observations lie between the vertical lines at 6 and 14 on the graph.

5. Raw Data to z-Scores.
To find the z-score of a data value, we must find how many standard deviations that data value is from the mean. If the mean is X and the standard deviation is σ, then a data value x has a z-score of

$$z = \frac{x - X}{\sigma}.$$

For this problem, we let $X = 50$ and $\sigma = 8$.

a. The data point 50 is the mean, so its z-score is 0. Check that the formula gives the same z-score.

b. A data value of 45 has a z-score of

$$z = (45 - 50)/8 = -5/8 = -0.625.$$

The data point 45 is 0.625 standard deviations *below* the mean.

c. A data value of 60 has a z-score of

$$z = (60 - 50)/8 = 10/8 = 1.25.$$

The data point 60 is 1.25 standard deviations *above* the mean.

d. A data value of 40 has a z-score of

$$z = (40 - 50)/8 = -10/8 = -1.25.$$

The data point 40 is 1.25 standard deviations *below* the mean.

e. A data value of 65 has a z-score of

$$z = (65 - 50)/8 = 15/8 = 1.875.$$

The data point 65 is 1.875 standard deviations *above* the mean.

f. A data value of 42 has a z-score of

$$z = (42 - 50)/8 = -8/8 = -1.0.$$

The data point 42 is 1.0 standard deviation *below* the mean.

g. A data value of 58 has a z-score of

$$z = (58 - 50)/8 = 8/8 = 1.0.$$

The data point 58 is 1.0 standard deviation *above* the mean.

h. A data value of 75 has a z-score of

$$z = (75 - 50)/8 = 25/8 = 3.125.$$

The data point 75 is 3.125 standard deviations *above* the mean.

7. z-Scores and Percentiles.
All percentiles are rounded to the nearest whole percentile.

a. An observation 1 standard deviation above the mean has a z-score of $z = 1$. According to Table 9.7, this z-score corresponds to the 84th percentile.

b. An observation 1.5 standard deviations below the mean has a z-score of $z = -1.5$. According to Table 9.7, this z-score corresponds to the 7th percentile.

c. An observation 2 standard deviations below the mean has a z-score of $z = -2.0$. According to Table 9.7, this z-score corresponds to the 2nd percentile.

d. An observation 1.5 standard deviations above the mean has a z-score of $z = 1.5$. According to Table 9.7, this z-score corresponds to the 93rd percentile.

e. According to Table 9.7, the 94th percentile corresponds to a z-score of 1.6, so this observation is 1.6 standard deviations above the mean.

f. According to Table 9.7, the 6th percentile corresponds to a z-score of about -1.55, so this observation is 1.55 standard deviations below the mean.

9. GRE Scores.

a. With a mean of 497 and a standard deviation of 115, a score of 650 has a z-score of $(650 - 497)/115 = 1.33$. According to Table 9.7, a z-score of 1.33 corresponds to about the 90th percentile.

b. A score in the 95th percentile has a z-score of about 1.7. This means the score is 1.7 standard deviations or $1.7 \times 115 = 196$ points above the mean. So the score is $497 + 196 = 693$.

11. Close to Normal Income Distributions. An income of $27,000 in this distribution has a z-score of ($27,000 − $30,000)/$6000 = −0.5. By Table 9.7, this income is in the 31st percentile. An income of $33,000 in this distribution has a z-score of ($33,000 − $30,000)/$6000 = 0.5. So this income is in the 69th percentile. This means the 69% of the people in the study earn less than $33,000 and 31% of the people in the study earn less than $27,000, Therefore 69% − 31% = 38% of the people earn between $27,000 and $33,000.

13. Heart Rates.

a. A range of heart rates between 55 and 85 is 1 standard deviation on either side of the mean. Therefore about 68% of the classmates have a heart rate in this range.

b. A range of heart rates between 40 and 100 is 2 standard deviations on either side of the mean. Therefore about 95% of the classmates have a heart rate in this range.

c. A heart rate of 90 is $(90 - 70)/15 = 20/15 = 1.33$ standard deviations above the mean and has a z-score of 1.33. By Table 9.7, this heart rate is in the 90th percentile. A heart rate of 50 has a z-score of $(50 - 70)/15 = -20/15 = -1.33$. By Table 9.7, this heart rate is in the 10th percentile. Therefore, 90% − 10% = 80% of the classmates have a heart rate between 50 and 90.

d. A heart rate of 45 has a z-score of $(45 - 70)/15 = -25/15 = -1.67$. By Table 9.7, this heart rate is in the 5th percentile. Only 5% of the classmates have a lower heart rate.

d. A heart rate of 110 has a z-score of $(110 - 70)/15 = 40/15 = 2.67$. By Table 9.7, this heart rate is in the 99.6 percentile. About 99.6% of the classmates have a lower heart rate.

15. Sample and Population Proportions. The proportion of left-handed people in the *population* is $305/1348 = 0.226 = 22.6\%$. The proportion of left-handed people in the *sample* is $23/100 = 23\%$. The sample is representative of the population.

17. Estimating Population Proportions. The proportion of people in the *sample* who support the visiting team is $103/1320 = 0.078 = 7.8\%$. The best estimate of the number of people in the population who support the visiting team is $7.8\% \times 4500 = 351$.

19. Distribution of Sample Proportions. The sample proportions will be distributed according to a normal curve with a mean of 0.45 and a standard deviation of $1/(2\sqrt{n})$. Because there are $n = 900$ people in the samples, the standard deviation is $1/(2\sqrt{900}) = 1/(2 \times 30) = 1/60 = 0.017$. A normal curve with these properties in shown below. The vertical lines show the mean and the interval one standard deviation on either side of the mean.

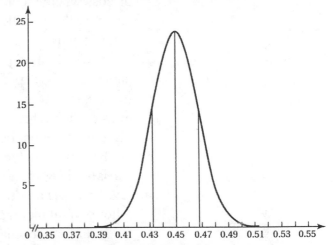

21. Election Predictions. The best estimate of the proportion of people supporting the mayor is $1300/2500 = 0.52 = 52\%$. The margin of error (for the 95% confidence interval) is $1/\sqrt{n} = 1/\sqrt{2500} = 1/50 = 0.02 = 2.0\%$. Thus there is a 95% likelihood that the percentage of voters supporting the mayor is between 50% and 54%. The 95% confidence interval is entirely above 50%, so you could conclude near certainty that the current mayor will win a majority of the votes.

23. Nielsen TV Ratings. The margin of error is $1/\sqrt{5000} = 0.014 = 1.4\%$. Thus the 95% confidence interval is 35% − 1.4% to 35% + 1.4%, or 33.6% to 36.4%.

Unit 9D Sample Issues in Statistical Research

Overview

This unit focuses on just four of countless case studies that illustrate statistics at work on contemporary issues. The topics that we present are

- SAT scores.
- the placebo effect,
- statistical significance, and
- the genetics of intelligence.

In the section on SAT scores, we look at trends in SAT scores in the last 50 years and ask whether the trends are representative of the educational performance of the entire population of high school students. We also discuss whether comparison of scores from one year to the next is justified and how the so-called recentering of SAT scores in 1996 affects comparisons.

A widely used practice in testing a new drug is to give half of the patients in the experiment a **placebo** — an ineffective substitute that looks like the real drug. It is well-known that patients often respond to the placebo even though it has no active ingredients. We discuss the impact of the placebo effect on the statistical analysis of such drug trials.

Following naturally the discussion of the placebo effect is the subject of **statistical significance**. Suppose that some patients in a study are given regular doses of Vitamin C, and the remaining patients are given a placebo. After several months the patients are put into two categories: those who got colds during the study and those that did not get colds. The results can be summarized in a 2×2 contingency table that shows the patients with and without Vitamin C who did and did not get a cold. The question is whether Vitamin C was really effective in preventing colds. The goal is to confirm one of two hypotheses:

- Hypothesis: Vitamin C is more effective than a placebo at preventing colds.
- Null Hypothesis: Vitamin C is no more effective than a placebo at preventing colds.

We don't go into the technical details of hypothesis testing that are found in most statistics books. However we do discuss what it means to decide between the hypothesis and the null hypothesis, and to establish statistical significance.

The unit closes with a qualitative discussion on the controversial subject of IQ scores and the genetics of intelligence which has some significant statistical aspects.

Key Words and Phrases

placebo effect	statistical significance	contingency table
hypothesis	null hypothesis	

Key Concepts and Skills

- understand some of the issues in analyzing and comparing SAT scores.
- understand the issues that arise in using a placebo in drug tests.
- construct and analyze a 2×2 contingency table.
- understand the meaning of statistical significance and the role of the null hypothesis.
- understand the issues involved with the use of IQ scores.

Solutions Unit 9D

1. Recentered SAT Verbal Scores.

a. A score of 500 relative to a mean of 430 has a z-score of $(500 - 430)/100 = 0.7$. This score falls in the 76th percentile.

b. A score of 500 relative to a mean of 500 has a z-score of $(500 - 500)/100 = 0.0$. This score falls in the 50th percentile.

5. Contingency Table for Grades and Study.

a. The hypothesis for the study is that attending the study session is more likely to lead to an improvement in a student's grade than not attending the study session. The null hypothesis is that attending the study session has no more effect on grade improvement than not attending the study session.

b. If the null hypothesis is true, then attending or not attending the study session should make no difference. Thus, we would expect that $220/400 = 55\%$ of the students will improve their grade.

c. If the hypothesis is true, then we would expect the percentage of students attending the study session who improved their grade to be significantly greater than 55% and the percentage of students not attending the study session who improved their grade to be significantly less than 55%.

d. In fact, the table shows that $145/200 = 72.5\%$ of the students attending the study session improved their grade, while $75/200 = 37.5\%$ of the students not attending the study session improved their grade. We see that the percentage of students attending the study session who improved their grade is 17.5 percentage points greater than the 55% expected by the null hypothesis. The percentage of students not attending the study session who improved their grade is 17.5 percentage points less than 55% expected by the null hypothesis.

e. The percentage of students who did improve their grade with the study session is significantly greater than what would be expected by chance. Similarly the number of students who improved their grade without the study session is significantly less than what would be expected by chance. These differences are statistically significant and it is reasonable to claim that the study session had an effect on grade improvement.

7. Drinking and Driving Court Case.

a. The hypothesis for the study is men are more of a risk for drinking and driving than women. The null hypothesis is that men and women are equal drinking and driving risks.

b. If the null hypothesis is true, being a man or a woman should make no difference in whether the subject has been drinking in the last two hours. Thus, we would expect that $93/619 = 15.0\%$ of the people in the study have been drinking in the last two hours.

c. If the hypothesis is true, then we would expect the percentage of men who have been drinking in the last two hours should be greater than 15% and the percentage of women who have been drinking in the last two hours should be less than 15%.

d. The table shows that $77/481 = 16.0\%$ of men were drinking in the last two hours, while $16/138 = 11.5\%$ of women were drinking in the last two hours. The percentage of men drinking in the last two hours is 1.0 percentage points higher than predicted by the null hypothesis. The percentage of women drinking in the last two hours is 3.5 percentage points lower than predicted by the null hypothesis.

e. The percentages expected by chance are close to the actual percentages for both men and women, suggesting that the differences are not statistically significant. The Supreme Court ruled that the difference between the percentage of drinking men and the percentage of drinking women was not significant: there should be no distinction in drinking age based on gender.

10. MATHEMATICS AND THE ARTS

Overview

In this chapter we explore a very different application of mathematics, namely music and the fine arts. As you will see, the connections between mathematics and the arts go back to antiquity. However, with the development of digital music (CD players) and fractal art, the connections are also quite modern.

Because much of the chapter deals with concepts from geometry, we open with a unit on the fundamentals of geometry. This unit may be familiar to you from previous courses. We then devote one unit to music, one unit to classical painting, one unit to proportion as it appears in art and nature, and a final unit to the contemporary topic of fractals. This chapter should provide you with a new perspective and change of pace!

Unit 10A Fundamentals of Geometry

Overview

Because much of this chapter is based on ideas and concepts from geometry, it makes sense to do some review. If much of this unit is familiar, just sit back and enjoy it!

The unit begins by defining and giving examples of the basic concepts of geometry: **points**, **lines**, **planes**, and **angles**. We then move on to familiar two-dimensional objects (also called plane objects). The most basic objects are **circles** and **regular polygons** such as **triangles**, **squares**, **rectangles**, **pentagons**, and **hexagons**. A few formulae are given along the way. You should know, or be able to find quickly, formulae for the

- area of a circle,
- circumference of a circle,
- area of a triangle, and
- area of a **parallelogram** (which include squares and rectangles).

We give several practical examples of the uses of these formulae.

The next subject is solid objects in three dimensions. You will want to be familiar with Table 10.2 which has formulae for the

- surface area and volume of a **rectangular prism** (box),
- surface area and volume of a **cylinder** (soda can)
- surface area and volume of a **sphere**.

We give several examples of these formulae applied to *practical* problems.

The unit closes with a very useful and far-reaching section on scaling laws. We first discuss scale models, such as maps or architectural models. If we take an object and increase all of its dimensions by a factor of, say 10, then the **scale factor** is 10. The important result of this section is that

- areas scale with the square of the scale factor, and
- volumes scale with the cube the scale factor.

So if you were to enlarge yourself by a scale factor of two, your surface area would increase by a factor of $2^2 = 4$, and your volume (and weight) would increase by a factor of $2^3 = 8$.

These scaling laws explain a lot of interesting biological and physical phenomena We answer questions such as, why does crushed ice keep your drink colder than large ice cubes? why can flies walk on ceilings? why does the Moon have no volcanoes?

Key Words and Phrases

geometry	Euclidean geometry	point
line	plane	dimension
angle	vertex	right angle
straight angle	acute angle	obtuse angle
perpendicular	parallel	radius
diameter	polygon	regular polygon
circumference	perimeter	parallelogram
rectangular prism	cube	cylinder
sphere	scale factor	scaling laws
surface to volume ratio		

Key Concepts and Skills

- define and give examples of point, line, and plane.
- convert angle measurements to fractions of a circle and vice versa.
- determine the perimeter and area of common plane objects (circle, triangle, square, rectangle, parallelogram).
- determine the surface area and volume of common three-dimensional objects (cube, rectangular prism, cylinder, sphere).
- use geometrical ideas and formulae to solve practical problems.
- understand scaling laws and surface area to volume ratio arguments.

Solution Unit 10A

3.

a. 1/3 circle = 1/3 × 360° = 120°.

b. 1/12 circle = 1/12 × 360° = 30°.

c. 1/20 circle = 1/20 × 360° = 18°.

5. Fractions of Circles. A fraction of 360° is the same fraction of a full circle.

a. 1° = 1/360 of a circle.

b. 4° = 4/360 = 1/90 of a circle.

c. 15° = 15/360 = 1/24 of a circle.

d. 30° = 30/360 = 1/12 of a circle.

7. Angle Practice.

a. The unknown angle has a measure of 120° since two angles that form a straight line total 180°.

b. The angle opposite the 30° angle also has a measure of 30°. This means that the two angles adjacent to the 30° angle have a measure of 150° since they form straight lines with the 30° angle.

9. Perpendicular and Parallel. Through a given point on a line exactly one line can be drawn perpendicular to the first line. A second line through the point would intersect the first line at an angle other than a right angle.

11. Circle Practice. Recall that the circumference of a circle with radius r is $C = 2\pi r$ and the area is $A = \pi r^2$.

a. A circle with a radius of $r = 6$ meters has a circumference of $C = 2\pi r = 2\pi(6 \text{ meters}) = 37.7$ meters. The area of the circle is $A = \pi r^2 = \pi(6 \text{ meters})^2 = 113.1$ sq meters.

b. A circle with a radius of $r = 4$ km has a circumference of $C = 2\pi r = 2\pi(4 \text{ km}) = 25.1$ km. The area of the circle is $A = \pi r^2 = \pi(4 \text{ km})^2 = 50.3$ sq km.

c. A circle with a radius of $r = 25$ cm has a circumference of $C = 2\pi r = 2\pi(25 \text{ cm}) = 157.1$ km. The area of the circle is $A = \pi r^2 = \pi(25 \text{ cm})^2 = 1963.5$ sq cm.

13. Distance Measurement.

a. Drawing a straight line between the bus stop and the theater forms a right triangle whose horizontal leg is 4 blocks or 160 meters long and whose vertical leg is 3 blocks or 120 meters long. The hypotenuse, which is the path a bird takes has a length of
$$\sqrt{(160^2 + 120^2)} = \sqrt{40,000} = 200 \text{ meters}.$$

b. Any stair-step path along the streets must cover at least four blocks horizontally and three blocks vertically. This adds up to 7 blocks or 280 meters.

15. No Calculation Required. The lower square portion of both barns have the same area. Looking at the upper triangular parts of the barns, we see that the base of both triangles have the same length (30 feet). However, the height of the triangle on the left is 25 feet, whereas the height of the triangle on the right is 20 feet. So the area of the triangle on the right is less that that area of the triangle on the left. So the area of the left barn is greater than the area of the right barn.

17. Back Yard. The entire yard has an area of 10 meters × 20 meters = 200 square meters. The flower garden that will not be seeded has an area of 4 meters × 4 meters = 16 square meters. The area of the semi-circular patio with a radius of 5 meters that will not be seeded is (1/2) × π × (5 meters)^2 = 39.3 square meters. Therefore, the area of the yard that will be seeded is (200 − 16 − 39.3) square meters = 144.7 square meters.

Solid Problems.

19. The volume of the convention hall is 40 meters × 60 meters × 10 meters = 24,000 cubic meters. Recall that 1 cubic meter = 10^6 cubic centimeters and that 1 liter = 1000 cubic centimeters. Therefore, the volume of the hall is

$$24,000 \text{ m}^3 \times \frac{10^6 \text{ cm}^3}{\text{m}^3} \times \frac{1 \text{ liter}}{1000 \text{ cm}^3} = 2.4 \times 10^7 \text{ liters}.$$

21. Recall that the volume of a circular cylinder is area of base × height = π × (radius)^2 × height. The area of the surface of a circular cylinder (excluding the circular end pieces) is 2× π× (radius) × (height). The duct has a length (height) of 25 feet and a radius of 10 in × (1 ft/12 in) = 0.833 ft. The volume is

$$\pi \times (0.833 \text{ ft})^2 \times 25 \text{ ft} = 54.50 \text{ ft}^3.$$

The surface area (and amount of paint needed), excluding the circular end pieces, is

$$2\times \pi \times (0.833 \text{ ft}) \times (25 \text{ ft}) = 130.85 \text{ ft}^2.$$

23. Recall that the volume of a sphere is (4/3) × π× (radius)^3 and the surface area is 4 × π× (radius)^2 . The radius of the ball is half of the diameter or 2.25 inches. Therefore, the volume of the lacrosse ball is

$$(4/3) \times \pi \times (2.25 \text{ in})^3 = 47.71 \text{ in}^3.$$

The surface area of the ball is 4 × π× (2.25 in)^2 = 63.62 in^2.

25. Architectural Model.

a. The height of the actual concert hall will be 25 times larger than the height of the scale model.

b. The surface area of the actual concert hall will be 25^2 = 625 times larger than the surface area of the scale model.

c. The volume of the actual concert hall will be 25^3 = 15,625 times larger than the volume of the scale model.

27. Tripling Your Size.

a. The scale factor in this problem is 1 to 3. Arm length is a linear dimension, so it increases by a factor of three when all of your linear dimensions are tripled.

b. Waist size is a linear dimension, so it increases by a factor of three when all of your linear dimensions are tripled.

c. The material in your clothes is proportional to surface area which increases by the square of the scale factor. So the amount of clothing needed increases by a factor of $3^2 = 9$ when all of your linear dimensions are tripled.

d. Weight is proportional to volume which increases by the cube of the scale factor. So your weight increases by a factor of $3^3 = 27$ when all of your linear dimensions are tripled.

29. Squirrels or People?

a. Squirrels have a much higher surface area to volume ratio than humans.

b. Squirrels lose much more heat for their volume (or weight) than humans. Therefore squirrels must maintain a higher rate of metabolism and eat more for their weight than humans.

31. Optimizing Area. Consider a square corral. The length of the sides of the corral is 132 meters/4 = 33 meters. So the area is 33 meters × 33 meters = 1089 square meters. Consider a circular corral. The radius of the corral is $132/(2\pi) = 21.0$ meters. The area of the corral is $\pi (21.0 \text{ m})^2 = 1387$ square meters. The maximum area is achieved by a circular corral.

35. Design of the Human Lung. Recall that the surface area of a sphere is $S = 4\pi r^2$ and the volume is $V = (4/3)\pi r^3$.

a. The radius of a single air sac is half the diameter or 1/6 mm. The surface area of a single air sac is

$$4\pi \times (1/6 \text{ mm})^2 = 0.35 \text{ mm}^2.$$

The volume of a single air sac is

$$(4/3)\pi \times (1/6 \text{ mm})^3 = 0.019 \text{ mm}^3.$$

Therefore, the total surface area of the air sacs is 3×10^8 air sacs × 0.35 mm²/air sac = 1.05×10^8 mm². The total volume of the air sacs is 3×10^8 air sacs × 0.019 mm³/air sac = 5.7×10^6 mm³.

b. If a single sphere had the same volume as the total volume of the air sacs (5.7×10^6 mm³), then its radius would satisfy the equation

$$5.7 \times 10^6 \text{ mm}^3 = (4/3)\,\pi \times r^3.$$

To find the radius of this sphere, multiply both sides of this equation by $(3/4\pi)$ which gives us

$$r^3 = 1.36 \times 10^6 \text{ mm}^3.$$

Now taking the cube root of both sides of this equation gives the radius $r = 111$ mm. The surface area of this sphere would be $4\pi \times (111 \text{ mm})^2 = 1.55 \times 10^5$ mm², which is much less than the total surface area of the air sacs.

c. If a single sphere had the same surface area as the total surface area of the air sacs (1.05×10^8 mm²), then its radius would satisfy the equation

$$1.05 \times 10^8 \text{ mm}^2 = 4\pi \times r^2.$$

To find the radius of this sphere, divide both sides of this equation by 4π giving

$$r^2 = 8.36 \times 10^6 \text{ mm}^2.$$

Taking the square root of both sides of this equation says that the radius of this sphere is 2.9×10^3 mm or 2.9 meters or almost 10 feet. The volume of this sphere would be 10^{11} mm³ which is far greater than the total volume of the air sacs.

d. The above calculations indicate that for their relatively small volume the air sacs have a tremendous surface area. The lung is designed to have a large surface area to facilitate air exchange.

37. The Chunnel. Each of the three tunnels is a half-cylinder with a volume of $(1/2) \times \pi r^2 h$. With a length of 50 kilometers (which will be the height of the cylinder) and a radius of 4 meters = 4×10^{-3} kilometers (which will be the radius of the cylinder), the volume of one tunnel is

$$(1/2) \times \pi \times (4 \times 10^{-3} \text{ km})^2 \times 50 \text{ km} = 1.26 \times 10^{-3} \text{ km}^3.$$

Therefore, the amount of Earth removed in building the three tunnels was about 0.004 cubic kilometers.

Unit 10B Mathematics and Music

Overview

The ties between mathematics and music go back to the ancient Greeks in about 500 B.C. The followers of Pythagoras discovered some of the basic laws that underlie our understanding of music today. They realized that the **pitch** of a musical note created by a plucked string is determined by the

frequency of the string — how many times the string vibrates each second. They also discovered that if the length of a string is halved, the frequency doubles, and the pitch goes up by an **octave**. With these few facts we can explain a lot.

In a standard scale that you might play on a piano or a guitar, one octave consists of 12 tones or **half-steps** (for example, the white and black keys between middle C and the next higher C). Here is the basic question we address: If we know the frequency of the first note of the scale, can we find the frequency of all 12 notes of the scale?

We show that to move up the scale a half step, we must *multiply* the frequency of the current tone by a fixed number. We give a brief argument showing that the magic number that generates the entire scale by multiplication is $f = 1.055946 \ldots$ or the twelfth root of 2. It turns out that the notes of the scale follow an exponential growth law (Chapter 7).

The ancient Greeks understood that when two notes have a pleasing sound when played together (**consonant tones**), then the ratio of their frequencies must be the ratio of two small numbers, such as 3/2 or 4/3. We investigate how these ratios of small numbers compare to the exact frequencies generated by the magic number f.

The unit closes with a few observations about the modern connections between music and mathematics: digital music, compact disks, and synthesizers.

Key Words and Phrases

sound wave	pitch	frequency
cycles per second	fundamental frequency	overtones
octave	scale	half-step
consonant tones	analog	digital

Key Concepts and Skills
- understand the relation between frequency and pitch.
- determine the frequency of notes separated by an octave.
- determine the frequency of notes on a 12-tone scale, given the frequency of the first note.
- understand the Greek's explanation for consonant tones in terms of ratios of small integers.
- understand the difference between analog and digital music.

Unit 10B Solutions

1. Octaves. Recall that to raise any tone by an octave, we must double (multiply by 2) its frequency. If we start with a tone that has a frequency of 110 cycles per second (cps), the tone an octave higher has a frequency of 2×110 cps = 220 cps. The tone two octaves above 110 cps has a frequency of 2×220 cps = 440 cps. The tones three and four octaves above 110 cps have frequencies of 880 cps and 1760 cps, respectively. Each jump of an octave corresponds to a doubling of the frequency.

3. Notes of a scale. Each half-step on a 12-tone scale corresponds to an increase in frequency by a factor of $f = 1.05946$. If we start at A with a frequency of 437 cps, then one half-step higher (A#) has a frequency of $437 \times 1.05946 = 463$ cps. Continuing in this way, we can produce the following table (rounding to the nearest whole number).

Note	Frequency (cps)
A	437
A#	463
B	491
C	520
C#	551
D	583
D#	618
E	655
F	694
F#	735
G	779
G#	825
A	874

5. Exponential Growth and Scales.

For this problem we can use the exponential growth law

$$Q = Q_0 \times 1.05946^n$$

where $Q_0 = 260$ cps is the starting frequency and Q is the frequency n half-steps above Q_0.

a. The frequency of the note $n = 5$ half-steps above C is
$$Q = 260 \times 1.05946^5 = 347 \text{ cps.}$$

b. According to Table 10.3, the interval of a fifth is $n = 7$ half-steps. The frequency of this note is
$$Q = 260 \times 1.05946^7 = 390 \text{ cps.}$$

c. An octave is 12 half-steps and a fourth is 5 half-steps. So the frequency of the note $n = 12 + 5 = 17$ half-steps above middle C is
$$Q = 260 \times 1.05946^{17} = 694 \text{ cps.}$$

d. The frequency of the note $n = 36$ half-steps above middle C is
$$Q = 260 \times 1.05946^{36} = 2080 \text{ cps.}$$

e. Four octaves is $4 \times 12 = 48$ half-steps. The frequency of the note $n = 48 + 3 = 51$ half-steps above middle C is
$$Q = 260 \times 1.05946^{51} = 4946 \text{ cps.}$$

7. Exponential Decay and Scales. To find the frequency of a note n half-steps above another note, we multiply by the factor 1.05946^n. Therefore to find the frequency of a note n half-steps *below* another note, we *divide* by the factor 1.05946^n. The frequency of the note $n = 5$ half-steps below the note with a frequency of 440 cps is $437/(1.05946^5) = 327$ cps. The frequency of the note $n = 8$ half-steps below the note with a frequency of 440 cps is $437/(1.05946^8) = 275$ cps.

9. Circle of Fifths.

a. Each half-step upward on the scale increases the frequency by a factor of $f = 2^{1/12} = 1.05946$. Raising a note by a fifth or seven half-steps, increases the frequency by a factor of $2^{7/12} = (1.05946)^7 = 1.498$.

b. Raising a tone by two fifths (14 half-steps) increases the frequency by a factor of $(1.05946)^{14} = 2.24$. Equivalently, raising a tone by two fifths increases the frequency by a factor $(1.498)^2 = 2.24$.

c. The table below shows the notes in the complete circle of fifths. We need 12 steps of a fifth to complete the circle which takes us through seven octaves. Each interval of a fifth raises the frequency by a factor of 1.498.

Note	Frequency (cps)
C	260
G	389
D	583
A	874
E	1309
B	1961
F#	2938
C#	4401
G#	6593
D#	9876
A#	14,794
F	22,162
C	33,198

d. The circle passes through 12 notes. Notice that the frequency of the last tone is almost 128 times the frequency of the first tone. Since $128 = 2^7$, this says that the frequency has doubled seven times which means that we have passed upward through seven octaves.

e. The frequency of the last tone is almost 128 times the frequency of the first tone.

f. Raising a note by a fourth (or five half-steps) increases its frequency by a factor of $2^{5/12} = (1.05946)^5$ or 1.335.

If we raise a tone by 12 intervals of a fourth, its frequency increases by a factor of $(2^{5/12})^{12} = 2^5$. This means that the circle of fourths extends over five octaves.

Unit 10C Perspective and Symmetry

Overview

It wasn't until the Renaissance (14th and 15th century) that painters attempted to draw three-dimensional objects realistically on a flat two-dimensional canvas. It took many years for these painters to perfect this technique, but in the end, they made a science of **perspective** drawing. In this unit we trace the development of perspective drawing and explore some of its mathematical necessities.

The key concept in perspective drawing is the **principal vanishing point**. It can be summarized as follows: If you are an artist standing behind a canvas, painting a real scene, then all lines that are parallel in the real scene and perpendicular to the canvas, must meet in a single point in the painting — this is the principal vanishing point. All other sets of parallel lines in the real scene (that are *not* perpendicular to the canvas) meet in their own vanishing points. All of the vanishing points (principal and otherwise) lie along a single line called the **horizon line**.

Another fundamental property of paintings and other objects of art is **symmetry**. Symmetry can mean many different things, but it often refers to a sense of balance. We define three different kinds of symmetry:

- **reflection symmetry**: an object can be "flipped" across a particular line and it remains unchanged (for example, the letter H).
- **rotation symmetry**: an object can be rotated through a particular angle and it remains unchanged (for example, the letter O).
- **translation symmetry**: an object or a pattern can be shifted, say to the right or the left, and it remains unchanged (for example, the pattern ....XXXXXX.... extended in both directions).

We investigate these symmetries in both geometrical objects and in actual paintings.

Symmetry arises in beautiful ways in **tilings** — patterns in which one or a few simple objects are used repeated to fill an region of the plane. We give several examples of how triangles and quadrilaterals can be translated and reflected to produce wonderful patterns. And you can try some tilings in the problems, too!

Key Words and Phrases

vanishing point	principal vanishing point	horizon line
symmetry	reflection symmetry	rotation symmetry
translation symmetry	tiling	

Key Concepts and Skills

- understand the role of the principal vanishing point in perspective drawing.
- draw simple objects in perspective using vanishing points.
- identify symmetries in simple objects.
- draw simple objects with given symmetries.
- create tilings from triangles or quadrilaterals using translations and reflections.

Solutions Unit 10C

1. Vanishing Points.

a. A vanishing point of the picture is that point at which the edges of the road appear to meet. According to the definition given in the book, this vanishing point is not the principal vanishing point. The principal vanishing point corresponds to all lines that are parallel in the real scene and perpendicular to the canvas. The road in this scene is not perpendicular to the canvas.

b. The lines connecting the tops of the telephone poles and bases of the telephone poles should also meet at the vanishing point.

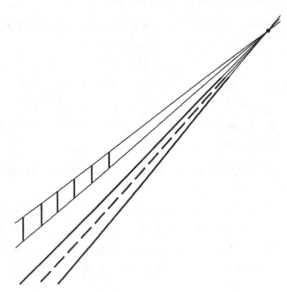

3. Drawing With Perspective.

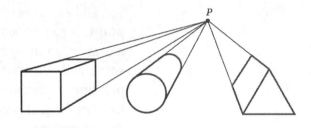

5. Proportion and Perspective.

a.

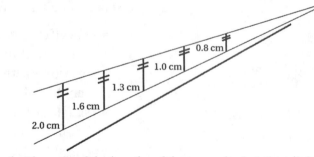

b. The ratio of the lengths of the two poles is 1.6 cm/2.0 cm = 0.8. This ratio must be maintained for all successive poles spaced 2 cm apart. So the third pole will have a height of 0.8 × 1.6 cm = 1.28 cm, the fourth pole will have a height of 0.8 × 1.28 cm = 1.024 cm, and the fifth pole will have a height of 0.8 × 1.024 cm = 0.82 cm.

c. If the poles are equally spaced in the drawing, then they would not be equally spaced in the real scene. Alternatively, if the poles are equally spaced in the real scene, then they would not appear equally spaced in the drawing.

9. Star Symmetries.

a. The four-pointed star has rotational symmetries of 90°, 180°, and 270°: it remains the same if it is rotated through any of these angles. It has four reflection symmetries: it can be reflected about a vertical line, a horizontal line, or either of two diagonal lines (all passing through the center of the star) and its appearance is unchanged.

b. The seven-pointed star has rotational symmetries of 360° ÷ 7 = 51.4° and all multiples of 51.4°, namely, 102.8°, 154.2°, 205.6°, 257.0°, 308.4°: it remains the same if it is rotated through any of these angles. There are seven reflection symmetries corresponding to lines through each of the seven points of the star.

Identifying Symmetries.

11. This figure has reflection symmetries: it can be reflected across a vertical line through its center, a horizontal line through its center, or either of its diagonals, and its appearance remains the same. It has rotation symmetries: it can be rotated though 90°, 180°, and 270°, and its appearance remains the same.

13. This figure (if imagined to be extended in both directions) has translation symmetry. It also has a reflection symmetry (across a horizontal line between the waves).

15. Tilings from Translating Triangles.

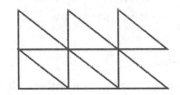

17. Tilings from Translating and Reflecting Triangles.

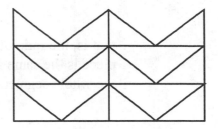

19. Tilings from Quadrilaterals.

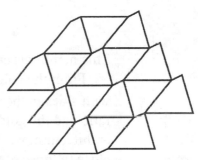

21. Why Quadrilateral Tilings Work. The angles around a point P are precisely the angles that appear inside of a single quadrilateral. Thus the angles around P have a sum of 360° and the quadrilaterals around P fit together perfectly.

Unit 10D Proportion and the Golden Ratio

Overview

In this unit we explore another fundamental aspect of art; that is proportion. As you will see, issues of proportion arise not only in the art created by humans, but in natural forms as well. The subject has a lot to do with aesthetics, our innate sense of what is beautiful. And one of earliest statements about proportion and beauty goes back to the ancient Greeks who introduced the golden ratio.

The first instance of the golden ratio arises in dividing a line segment. What division of a line segment has the most visual appeal and balance? More specifically, suppose that the line segment has a length of $L + 1$ units and we want to divide it into two pieces of length L and 1. How should we choose L? The Greeks answered that the best division is the one that makes

$$\frac{\text{length of long piece}}{\text{length of short piece}} = \frac{\text{length of entire piece}}{\text{length of long piece}},$$

that is, the ratio of the length of the long piece to the length of the short piece is the same as the ratio of the length of the long piece to the length of the whole segment. We can also write this as

$$\frac{L}{1} = \frac{L+1}{L}.$$

We show that the value of L that makes this happen is the special (irrational) number

$$\phi = \frac{1+\sqrt{5}}{2} = 1.61803\ldots$$

This number is called the **golden ratio** or **golden section**.

From the golden ratio, we can define a **golden rectangle**. Any rectangle with sides in the ration of ϕ is called a golden rectangle. Both the golden ratio and the golden rectangle arise throughout the history of art. Great architectural works (for example, the Greek Parthenon) have dimensions close to those of the golden rectangle (although some claim that these examples are coincidences). Many common objects such as post cards and cereal boxes also have dimensions of golden rectangles. We also look at some examples of how the golden rectangle appears in nature.

There is one final connection that is too intriguing to ignore. We introduce the **Fibonacci sequence**, which was first used as a population model in the 13th century. Each term of the sequence is formed by adding the two previous terms:

$$1, 1, 2, 3, 5, 8, 13, 21, 34, \ldots\ldots$$

We show how this sequence is related, perhaps unexpectedly, to the golden ratio. And the circle is closed by observing that the Fibonacci sequence also appears in the art work of nature. This is a unit with several seemingly unrelated ideas that eventually become linked in a beautiful way.

Key Words and Phrases

golden ratio	golden rectangle	logarithmic spiral
symmetry	Fibonacci sequence	

Key Concepts and Skills
- understand the golden ratio as a proportion and divide a line segment according to the golden ratio.
- construct and identify golden rectangles.
- generate the Fibonacci sequence and find the ratio of successive terms.
- understand the connection between the golden ratio and the Fibonacci sequence.

Solutions Unit 10D

Throughout these solutions, we will use a value of $\phi = 1.62$ for the Golden Ratio.

1. Golden Ratio. Recall that the Golden Ratio is the ratio of the length of the whole line (4 inches) to the length of the longer segment *and* the ratio of the longer segment to the length of the shorter segment. Thus, the longer segment should have a length of (4 inches)/1.62 = 2.47 inches and the shorter segment should have a length of (2.47 inches)/1.62 = 1.52 inches. Notice that the lengths of the long and short segments sum to 4.0 inches, as they should.

Dimensions of Golden Rectangles.

3. If the short side of a Golden Rectangle has a length of 1.5 inches, then the long side must have a length of

$\qquad$ 1.5 inches $\times \phi = 1.5$ inches $\times 1.62 = 2.43$ inches.

If the long side of a Golden Rectangle has a length of 1.5 inches, then the short side must have a length of

$\qquad$ 1.5 inches $\div \phi = 1.5$ inches $\div 1.62 = 0.93$ inches.

In both cases the ratio of the side lengths is ϕ.

5. If the short side of a Golden Rectangle has a length of 6.4 km, then the long side must have a length of

$\qquad$ 6.4 km $\times \phi = 6.4$ km $\times 1.62 = 10.37$ km.

If the long side of a Golden Rectangle has a length of 6.4 km, then the short side must have a length of

$\qquad$ 6.4 km $\div \phi = 6.4$ km $\div 1.62 = 3.95$ km.

In both cases the ratio of the side lengths is ϕ.

9. Properties of ϕ.

b. You can verify that $\phi^2 = (1.62)^2 = 2.62 = \phi + 1$.

11. The Lucas Sequence.

a. The relation for the Lucas numbers gives the following table for the first ten numbers.

n	L_n	L_n / L_{n-1}
1	1	–
2	3	3
3	4	1.33
4	7	1.75
5	11	1.57
6	18	1.64
7	29	1.61
8	47	1.62
9	76	1.617
10	123	1.618

b. Looking at the above table, the ratios approach ϕ.

Unit 10E Fractal Geometry

Overview

In this unit we explore perhaps the more recent mathematical development discussed in this book. **Fractal geometry** arises with the observation that classical geometry (invented by the ancient Greeks and studied by all of us in high school) works very well for regular objects such as circles and square. But many objects in the real world, particularly natural forms, are far more complex and are not described well by classical geometry. Fractal geometry was proposed, in part, to describe the complicated forms we find in nature.

We being by looking at what happens when we measure the length of a line segment. If we continually magnify the line we don't see anything new. This means that if we measure the length of a line segment, we get the same length regardless of the length of the ruler. Similarly, if we measure the area of a square, we get the same area regardless of how small the ruler might be. And if we measure the volume of a cube, we get the same volume regardless of how small the ruler is. This leads to the conclusion that a line is a one-dimensional object, a square is a two-dimensional object, and a cube is a three-dimensional object.

Here is how we define the **dimension** of an object. We imagine successively reducing the length of our ruler by a **reduction factor** R. Each time we reduce the length of the ruler, we observe by what factor the number of **elements** increases. For regular objects, we find that

- For a one-dimensional object (e.g., a line segment), $N = R^1$.
- For a two-dimensional object (e.g., a square), $N = R^2$.
- For a three-dimensional object (e.g., a cube), $N = R^3$.

In each case, we see that $N = R^D$, and D is the **dimension** of the object.

For irregular objects, we result may be quite different. If we carry out the same measurement process, we may find that the relationship between R and N has the form $N = R^D$, but D is no longer a whole number. If D is not a whole number, the object is called a **fractal**.

With this definition in hand, we look at some examples of fractal objects, notably the **snowflake curve,** the **Cantor set,** the **Sierpinski triangle**, and the **Sierpinski sponge**. We also discuss how to measure the fractal dimension of realistic objects such as coastlines. We close with the most famous fractal of all, the **Mandelbrot set**.

Key Words and Phrases

fractal geometry	element	reduction factor

fractal dimension   snowflake curve   snowflake island
self-similar   Sierpinski triangle Sierpinski sponge
Cantor set   Mandelbrot set

Key Concepts and Skills

- understand the meaning of dimension for regular objects (line, square, and cube).
- understand the meaning of dimension for fractal objects.
- determine the dimension of simple fractal objects.

Solutions Unit 10E

P 1. Ordinary Dimensions for Ordinary Objects.

a. Start with a meter stick and measure the length of the sidewalk, counting the number of times you lay the stick down on the sidewalk. Now reduce the length of the stick to 0.1 meter — a reduction by a factor of $R = 10$. Repeat the measurement of the sidewalk and count the number of times you lay the stick down on the sidewalk. You should find that the number of elements for this second measuring is $N = 10$ times the number of elements of the first measurement. If this pattern continues, that every time the ruler is decreased in length by a factor of $R = 10$, the number of elements increases by a factor of $N = 10$, then the sidewalk is a one-dimensional object.

3. Ordinary and Fractal Dimensions.
Recall that R is the factor by which the length of the ruler is reduced (reduction factor) and N is the corresponding factor by which the number of elements increases.

a. In this case $N = R^1$, so the dimension is 1 and the object is ordinary (non-fractal).

b. In this case $N = R^2$, so the dimension is 2 and the object is ordinary (non-fractal).

c. In this case $N = R^3$, so the dimension is 3 and the object is ordinary (non-fractal).

d. Using the general expression $N = R^D$, we see that for this object, $4 = 3^D$. Following the procedure of Example 1, we see that $D = \log_{10} 4/\log_{10} 3 = 1.26$. The object is a fractal.

e. Using the general expression $N = R^D$, we see that for this object, $12 = 3^D$. Following the procedure of

Example 1, we see that $D = \log_{10} 12/\log_{10} 3 = 2.26$. The object is a fractal.

f. Using the general expression $N = R^D$, we see that for this object, $36 = 3^D$. Following the procedure of Example 1, we see that $D = \log_{10} 36/\log_{10} 3 = 3.26$. The object is a fractal.

7. The Cantor Set.
At each stage of the construction of the Cantor Set, the middle third is removed from every line segment. When the ruler has its original length, it will detect one element. When the ruler is 1/3 its original length, it will detect two elements. When the ruler is 1/9 its original length, it will detect four elements. The number of elements doubles from 1, 2, 4, 8, to 16, ... each time the ruler is reduced in length by a factor of three. Thus, $N = 2$ and $R = 3$. The fractal dimension is

$$D = (\log_{10} 2)/(\log_{10} 3) = 0.301/0.477 = 0.63.$$

9. Fractal Dimension from Measurements.

a. The table shows the original data and the values of log r and log L.

The Coastline of Dragon Island						
r meters	100	10	1	0.1	0.01	0.001
L meters	315	1256	5000	19905	79244	315,479
$\log_{10} r$	2	1	0	−1	−2	−3
$\log_{10} L$	2.5	3.1	3.7	4.3	4.9	5.5

b, c. The graph is very close to a straight line, which indicates that the coastline is a self-similar fractal.

d. The slope of the (log r, log L) relation is $1 - D$, where D is the fractal dimension. Choosing several pairs of points and computing the slope, we see that the slope of the line is about −0.6 (for example, $(3.1 - 2.5)/(1 - 2) = -0.6$). Therefore, the fractal dimension of the coastline is

$$D = 1 - \text{slope} = 1 - (-0.6) = 1.6.$$

11. DISCRETE MATHEMATICS IN BUSINESS AND SOCIETY

Overview

In this chapter we will explore a different area of mathematics and many different kinds of applications. Discrete mathematics is the area of mathematics that deals with separate or individual items. The items might be people, objects, choices, or geographical locations. We have already seen some discrete mathematics in this book: the counting methods presented in Unit 8A are examples of discrete mathematics. This chapter will look at two more topics from discrete mathematics.

In Units 11A, B, and C, we will look at applications of network theory. As you will see, a network is a mathematical model that illustrates how the various components of a system are interrelated. For example, the power lines that connect several communities or the trading relations between several countries can be displayed using a network. The applications of network theory are endless. The mathematics is quite different because it is visual and involves very little computation. The second topic of the chapter, covered in Units 11D and E, is voting systems. In these units, you will see how unexpected complications and paradoxes arise when there are more then two candidates in an election. We also look at some of the real voting issues that arise in American politics. It is an extremely practical chapter, filled with new ideas.

Unit 11A Network Analysis

Overview

In this unit we introduce the idea of a **network** (called a **graph** in some books). A network is a collection of points (called **vertices**) connected by lines or **edges**. A nice example of a network arises in the 18th century **Konigsberg bridge problem** that led to the invention of network theory. We also give several more practical situations that lead to networks.

One of the fundamental network questions is whether it is possible to find a path that traverses every edge of the network exactly once and returns to the starting vertex. If such a path exists, it is called an **Euler circuit** (pronounced *oiler*, in honor of the mathematician who invented network theory). An Euler circuit would be of interest to an inspector, a meter reader, or a delivery person, who would like to do his/her job in the most efficient way. It turns out that there is a simple rule that determines whether a network has an Euler circuit.

> an Euler circuit exists for a network only if each vertex has an even number of edges attached to it.

Having determined whether a network has an Euler circuit doesn't tell you how to find it! So we also provide a rule, called the **burning bridges rule**, that tells you how to find an Euler circuit. The rule says that

> you may begin your circuit from any vertex in the network. However, as you choose edges to follow for your path, avoid using an edge that is the *only* connection to a part of the network that you have not already visited.

There is a lot of terminology associated with networks, some of which we introduce at this point. Here are a few important terms.

- a path within a network that begins and ends at the same vertex is a **circuit**.

- a **cycle** is a network in the shape of a closed ring in which each vertex is connected to exactly two other vertices.

- a network is **complete** if every vertex is connected to every other vertex.

- a **tree** is a network in which all of the vertices are connected, but there are no cycles (it looks like a tree with roots and branches).

- the **order** of a network is the number of vertices.

- the **degree** of a vertex is the number of edges connected to it (each vertex in a network could have a different order).

The last problem in this unit deals with networks in which every edge has a number or **weight** associated with it. The weights often represent distances or costs. A **spanning network** for such a network is a set of edges within the network that connects every vertex to every other vertex. In other words, if you walk along a spanning tree you can reach every vertex from every other vertex. A given network could have several different spanning networks, so a question of great importance is which, of many different spanning networks, has the minimum cost? This is called the **minimum cost spanning network** problem. (You might want to convince yourself that the minimum cost spanning network must be a tree).

We give some practical examples of minimum cost spanning tree problems and then provide an efficient method, called **Kruskal's algorithm**, for finding the minimum cost spanning tree of a network.

Key Words and Phrases

network	edge	vertex
Euler circuit	burning bridges rule	cycle
circuit	tree	order
degree	complete	spanning network
minimum cost spanning network	Kruskal's algorithm	

Key Concepts and Skills

- draw a network as a model of a given practical situation.
- determine if a given network has an Euler circuit and find one.
- understand the applications of Euler circuits.
- classify a network (in terms of cycles, tree, complete), identify the order of the network and the degree of each vertex.
- understand the applications of minimum cost spanning networks.
- apply Kruskal's algorithm to find the minimum cost spanning network.

Solutions Unit 11A

P 1. Discrete or Continuous?

a. Buses arrive at distinct times separated by periods in which no buses arrive. This is a <u>discrete</u> process.

b. The path of a skydiver can be viewed as a smooth curve, with no jumps or gaps, along which her altitude changes *continuously*. This is a <u>continuous</u> process.

5. City Streets.

a. Letting vertices represent intersections and edges represent *streets*, the network for the map is shown below. This network does not have an Euler circuit because there are vertices with an odd degree.

b. Letting vertices represent intersections and edges

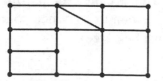

represent *sidewalks*, the network for the map would

look the same as below except that every edge would be replaced by *two* edges (because every street has two sidewalks, one on each side). The network for sidewalks does have Euler circuits because every vertex has an even degree.

7. Friendships and International Trade. We can use vertices to represent the four people, Amy, Beth, Cate, and Daniel, and edges to represent the reciprocal relation "trades with." The resulting network is shown below. As with all networks, the precise shape makes no difference. The important feature is the connections between the vertices. A similar network could be used to represent countries with edges standing for the relation "is a trading partner of." The network would display the trading relations between the countries.

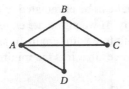

a.

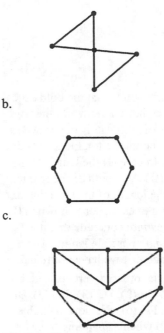

9. Euler Circuits.

a. We can let vertices represent the town and the islands and we can let edges represent ferry routes. So we need four vertices and five edges.

b. The network for the town is shown below.

b.

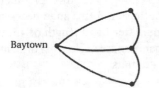

Baytown

c. Since there is at least one vertex with an odd number of edges attached to it, the network has no Euler circuit.

c.

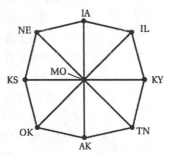

11. Checking Parking Meters. Letting vertices represent intersections and edges represent sidewalks, the network for the map is shown below. Every vertex has an even degree which tells us that an Euler circuit can be found. In fact, there are many Euler circuits.

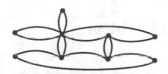

23. Neighboring States. The Missouri network is shown below. Note that there is a vertex for each state and vertices are connected with an edge whenever the corresponding states are neighbors. The order of the network is nine (there are nine vertices) and the degree of the Missouri vertex is eight since Missouri has eight neighbors.

Networks Terminology.

13. (a) The network has order 8 (a total of 8 vertices), (b) vertices A, D, E, F, G, H have degree 3, vertices B and C have degree 4, (c) the network has no special form, and (d) with vertices of odd degree, there are no Euler circuits.

15. (a) The network has order 8 (a total of 8 vertices), (b) all vertices have degree 4, (c) the network has no special form (actually it is called *bipartite* since every vertex of the top set of four vertices is connected to every vertex of the bottom set of four vertices), and (d) because all vertices have even degree, there are Euler circuits.

17. (a) The network has order 5, (b) vertex A, has degree 1, vertices C and D have degree 2, vertex E has degree 3, vertex B has degree 4, (c) the network has no special form, and (d) with vertices of odd degree, there are no Euler circuits.

19. Drawing Networks. Example of networks with the given properties are shown below. They are not unique!

IA
NE   IL
    MO
KS   KY
OK   TN
    AK

25. Soccer Tournament. The network for the soccer tournament, shown below, has five vertices, one for each team; thus the network has order five. Vertices A, C, D, and E have degree three (those teams have three games), while vertex B has degree two. The network is not complete (every vertex is not connected to every other vertex) because each team does not play every other team. If there were six teams in the tournament, then each team could play three teams in the remaining three weeks.

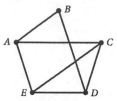

27. Spanning Networks. First notice that the bold edges really do form a spanning tree in each network: the bold edges insure that every vertex is ultimately connected to every other vertex. We must count the total length of the bold edges in each network. In case (I) the length of the spanning tree is 24, in case (II) the length of the spanning tree is 29, and in case (III) the length of the spanning tree is 31. Of these three spanning trees, the one shown in (I) has the shortest length. We *cannot* conclude that it is spanning tree with the *minimum* possible length. There may be a shorter spanning that we have not found.

29. Kruskal's Algorithm. The ordered vertices and their lengths are: AC (1), AB (2), CE (2), DE (3), EG (3), BD (4), CD (4), CF (5), DG (5), FG (6). We now assemble these edges in order and stop when a spanning tree is reached. If we assemble the first eight edges, AC, AB, CE, DE, EG, BD, CD, and CF, then every vertex is connected to every other vertex. However, there are also some unnecessary loops that can be eliminated without losing the spanning property. If we eliminate edges BD and CD, we have a spanning tree with length 16 and Kruskal's algorithm guarantees that it is the minimum cost spanning tree.

31. Local Area Networks. The goal is to find the minimum cost spanning tree for this network. Using Kruskal's algorithm, we first order the edges in increasing order: $c(8), f(8), i(9), b(10), e(10), j(10), g(11), h(11), d(12), a(15)$. We can now assemble them in order and stop when all of the vertices are connected. The edges, c, f, i, b, e, j make a spanning tree, however either edge b or e can be removed because it forms an unnecessary loop. The resulting spanning tree has a length of 45. It is the minimum length spanning tree. The cost of connecting all the workstations is 45 units.

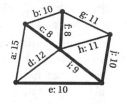

Unit 11B The Traveling Salesman Problem

Overview

This unit is devoted to one of the most famous of all network problems. However the **traveling salesman problem** has many applications beyond its namesake. Consider a network in which the vertices might be the cities that a salesman must visit and the edges are the airline routes that connect them along with their distances. A very practical problem is to find the route of minimum length that visits every city exactly once and returns to the starting city. This is an example of a traveling salesman problem.

We begin the discussion by introducing a preliminary idea: a **Hamiltonian circuit** is a path through a network that visits every vertex exactly once and returns to the starting vertex. While a Hamiltonian circuit passes through every vertex, it may not traverse every edge (as an Euler circuit does). Not every network has a Hamiltonian circuit, and furthermore, it is very difficult to determine whether a general network has a Hamiltonian circuit. In fact, there is no simple rule (as there is with Euler circuits) that determines when a network has a Hamiltonian circuit.

There are a few cases in which it is possible to find Hamiltonian circuits easily. If a network is complete (every vertex is connected to every other vertex), then Hamiltonian circuits abound. In fact there are $(n - 1)!/2$

Hamiltonian circuits in a complete network of order n. The number of Hamiltonian circuits increases explosively with the order of the network.

We must now imagine a network (which may or may not be complete) in which all of the edges have distances on them. To solve the traveling salesman problem for this network, we must find not just any Hamiltonian circuit, but the one with the shortest length. This turns out to be an incredibly difficult problem, especially for problems of practical interest in which there may be thousands of vertices in the network. We describe a method, called the **nearest neighbor method**, that can be used to find nearly optimal traveling salesman solutions.

We close the unit with some general remarks about a branch of mathematics called **operations research** that includes problems such as the traveling salesman problem. Operations research often handles problems from business and management that involve finding the most efficient or least expensive way to carry out a task. Airlines and manufacturing companies use operations research every day to streamline their operations, and our lives are affected by this area of mathematics continuously.

Key Words and Phrases

traveling salesman problem	Hamiltonian circuit	nearest neighbor method
operations research		

Key Concepts and Skills

- understand the applications of Hamiltonian circuits.
- determine if a given circuit is a Hamiltonian circuit for a network.
- find the number of Hamiltonian circuits in a complete network.
- understand the applications of the traveling salesman problem.
- apply the nearest neighbor method to find a traveling salesman solution.

Unit 11B Solutions

1. Number of Hamiltonian Circuits. Recall that there are $(n-1)!/2$ Hamiltonian circuits in a complete network of order n.

a. In a complete network of order 8 there are $7!/2 = 5040/2 = 2520$ Hamiltonian circuits.

b. In a complete network of order 18 there are $17!/2$ or approximately 1.78×10^{14} Hamiltonian circuits.

c. There are 60 seconds in a minute and 31,536,000 seconds in a year. Checking all the Hamiltonian circuits in a complete network of order 8 would require 2520 seconds = 42 minutes. Checking all the Hamiltonian circuits in a complete network of order 18 would require 1.78×10^{14} sec $\times$ (1 yr/31,536,000 sec) = 5,644,343 years.

Practice with Hamiltonian Circuits. Recall that given a network, a Hamiltonian circuit must visit all of the vertices once and return to the starting point.

3. The circuit shown is not a Hamiltonian circuit because it doesn't return to the starting point. It can be made into a Hamiltonian circuit by moving the vertical edge so it returns to the leftmost vertex.

5. The circuit shown is not a Hamiltonian circuit because it doesn't return to the starting point. It cannot be made into a Hamiltonian circuit.

7. Traveling the National Parks.

a. The network will look exactly like Figure 11.29 of the text except that the vertex for Zion will be missing along with all of the edges connected to it. The resulting network has four vertices and a total of six edges.

b. The leg from Bryce to Canyonlands is 136 miles, Canyonlands to Capitol Reef is 75 miles, Capitol Reef to Grand Canyon is 151 miles for a total of 362 miles. With the return to Bryce, the trip has a length of 470 miles.

c. Starting at Grand Canyon, the nearest neighbor method would send us to Bryce first at a distance of 108 miles. The next stop would be Capitol Reef, 69 miles away. The next stop would be Canyonlands , 75 miles away. To return to Grand Canyon, we need to travel another 202 miles. The total length is 454 miles.

d. Starting at Bryce, the nearest neighbor method would first send us to Capitol Reef, 69 miles away. The next stop would be Canyonlands , 75 miles away. The next stop would be Grand Canyon, 202 miles away. To return to Bryce, we need to travel another 108 miles. The total length is 454 miles. This is the same circuit that results from starting at the Grand Canyon as in part (c).

e. There are three different Hamiltonian circuits possible in this network of order four. They have lengths 454, 470, and 558. So we have found the Hamiltonian circuit with the shortest length in parts (c) and (d).

9. A Traveling Salesperson Problem. Notice that since the network is not complete (every vertex is not connected to every other vertex), it may not always be possible to apply the nearest neighbor method depending on the starting vertex. For example, the nearest neighbor method does find the circuit FEABCDF which has length 23. The circuit BCDFEAB is not produced by the nearest neighbor method (rather by trial and error) and it also has length 23. These two circuits *appear* to be the shortest possible.

11. Car Shuttles. There are three different Hamiltonian circuits in this network. The nearest neighbor method gives the circuit ACBDA which has length 14. The other two possible circuits, ACDBA with length 12 and ABCDA with length 16, can be found by trial and error. The shortest route for Abe is ACDBA. All of the drivers would use the same route.

13. Overnight Delivery. The time savings with the new route is 890 hours − 860 hours = 30 hours. At $15,000 per hour, this savings amounts to 30 hours × $15,000/hour = $450,000. In a year, the savings would be 365 days/year × $450,000/day = $164,250,000.

15. Distribution of Rental Cars. Move 10 cars from the first oversupply site to the first demand site; this costs 10 × $10 = $100 and leaves the first oversupply site empty. Then move 5 cars from the second oversupply site to the first demand site; this costs 5 × $30 = $150 and fills the first demand site. Finally move 15 cars from the second oversupply site to the second demand site; this costs 15 × $20 = $300. The total cost is $100 + $150 + $300 = $550.

Unit 11C Project Design

Overview

In this unit we look at a very different use of networks that is just as practical as all the previous applications. Imagine a large project, such as building a house or planning a shopping center(or even landing a person on the Moon), that consist of many smaller tasks. Some of the smaller tasks can be done at the same time, but some must be done before or after others. For example, in a house building project, you must build the walls before you can paint the walls.

As you will see in this unit, a network can be used to schedule the smaller tasks and illustrate which tasks must precede other tasks. The network also shows the estimated time for the completion of each task. Once the scheduling network is created, several important questions can be asked. For example, we can determine the minimum time for completion of the project. The minimum completion time corresponds to the *longest* path through the network; this longest path is called the **critical path**. The critical path consists of all the **limiting tasks** for the project.

We can also determine when each task can be started and finished. The following four quantities are useful for this purpose:

- **earliest start time** for a task is the soonest that a task can be started after the beginning of the project.
- **earliest finish time** for a task is the soonest that a task can be finished after the beginning of the project.
- **latest start time** for a task is the latest that a task can be started after the beginning of the project and still have the project finish on time .
- **latest finish time** for a task is the latest that a task can be finished after the beginning of the project and still have the project finish on time .

A general rule is that you should start at the *beginning* of the project and work *forward* through the network to find the earliest start time and earliest finish time. You should start at the *end* of the project and work *backward* through the network to find the latest start time and latest finish time.

Finally, for each task we define the **slack time** as the difference between the latest start time and the earliest start time (or equivalently as the difference between the latest finish time and the earliest finish time). Tasks with a zero slack time are on the critical path. Tasks with a non-zero slack time are not on the critical path and can be delayed without affecting the outcome of the entire project. Knowing the slack times allows a project manager to decide which tasks must be completed on time and which tasks can be postponed.

The entire unit really consists of one detailed example that illustrates all of these points. After that you are ready to try some problems and use these ideas to organize your life!

Key Words and Phrases

critical path	limiting task	earliest start time
earliest finish time	latest start time	latest finish time
slack time		

Key Concepts and Skills

- create and interpret a network as a schedule for a multi-stage project.

• find the critical path for a scheduling network.
• find the earliest start time, earliest finish time, latest start time, latest finish time, and slack times for all task in a scheduling network.

Solutions Unit 11C

Scheduling a Paint Job.

1. The longest path between A and E passes along tasks a, e, and f.

3. The longest path through the entire network passes along tasks a, e, f, g, i, and k (vertices A, B, D, E, F, G, H). This is the critical path.

5. The earliest time that task d can start is 1 hour; the latest time that task d can start is 2 hours; it is not on the critical path. The earliest time that task f can start is 2.5 hours; the latest time that task f can start is 2.5 hours; it is on the critical path.

7. The earliest time that task d can finish is 1.5 hours; the latest time that task d can finish is 2.5 hours; it is not on the critical path. The earliest time that task f can finish is 4.5 hours; the latest time that task f can finish is 4.5 hours; it is on the critical path.

9. Because task c is not on the critical path and has 1.5 hours of slack time, a one hour delay in this task will not delay the project.

11.

Task	EST	LST	EFT	LFT	Slack
a	0.0	0.0	2.0	2.0	0
b	0.0	1.0	1.0	2.0	1.0
c	0.0	1.5	1.0	2.5	1.5
d	1.0	2.0	1.5	2.5	1.0
e	2.0	2.0	2.5	2.5	0
f	2.5	2.5	4.5	4.5	0
g	4.5	4.5	7.5	7.5	0
h	4.5	5.5	6.5	7.5	1.0
i	7.5	7.5	9.5	9.5	0
j	9.5	10.0	10.0	10.5	0.5
k	9.5	9.5	10.5	10.5	0

Building a Hotel.

13. The longest path between B and G passes along tasks e, f, h, and i.

15. The length of the critical path is 19 months.

17. The earliest time that task d can start is 3 months; the latest time that task d can start is 5 months; it is not on the critical path. The earliest time that task h can start is 7 months; the latest time that task h can start is 7 months; it is on the critical path.

19. The earliest time that task d can finish is 5 months; the latest time that task d can finish is 7 months; it is not on the critical path. The earliest time that task h can finish is 15 months; the latest time that task h can finish is 15 months; it is on the critical path.

21. Because task e is on the critical path, a one month delay in this task will delay the project.

Doing the Laundry.

23. The slack time for a task is LST – EST or LFT – EFT. For this job, the slack times are: a 0, b 0, c 10, d 0, e 15, f 30, g 0, h 0.

25. The length of the critical path is 150 minutes.

27.

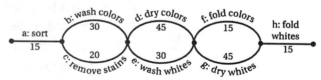

Cooking Pasta Dinner.

29. The critical path passes through the tasks with no slack time. In this case the critical path is a, c, and f.

31. Tasks (a, b) and (c, d, e) take place at the same time. Task f must be done alone.

Unit 11D Voting: Does the Majority Always Rule?

Overview

In this and the following unit we investigate mathematical problems associated with voting. This may sound like an unusual application of mathematics, but voting problems have been studied for several centuries and it has long been known that curious things can happen in voting systems. We will study such curious things in these two units.

The discussion begins with elections between two candidates. Throughout these units a candidate can be interpreted as a choice between two or more alternatives. For example, a candidate may be a person running for office or a brand of bagels in a taste test. With only two candidates, the rules are straightforward: the **majority** rules. This means that the candidate with the most votes (which must be more than 50% of the vote) wins the election.

However, even with majority rule, there are some interesting situations that can arise. We first look at presidential elections in which the winner is chosen, not by the **popular vote**, but by the **electoral vote**. Historically, there have been U.S. presidential elections in which a candidate won the popular vote, but lost the election.

We then look at variations on majority rule that often involve **super majorities**. For example, many votes in the U.S. government require more than a 50% majority: it takes a 2/3 super majority in both houses of Congress, followed by a 3/4 super majority vote of the states to amend the U.S. Constitution. More than a 50% majority of a jury is required to reach a verdict in a criminal trial.

Things get interesting when we turn to elections with three or more candidates. Often such elections are based on a **preference schedule** in which each voter ranks the candidates in order of preference. For example, the following preference schedule shows the outcome of an election among five candidates that we call A, B, C, D, and E.

First	A	B	C	D	E	E
Second	D	E	B	C	B	C
Third	E	D	E	E	D	D
Fourth	C	C	D	B	C	B
Fifth	B	A	A	A	A	A
	18	12	10	9	4	2

There was a total of 55 voters and 18 voters ranked A first, D second, E third, C fourth, and B last. The other columns are interpreted in a similar way. The question is: how do we determine a winner of the election?

The remainder of the unit presents five methods to determine a winner to an election with a preference schedule.

- **plurality**: the candidate with the most votes wins (candidate A would win the above election).

- **top two runoff**: the top two candidates have a runoff in which the votes of the losing candidates are redistributed to the top two candidates (candidate B would win the above election).

- **sequential runoff**: the candidates with the fewest first place votes are successively eliminated one at a time, votes are redistributed, and runoff elections are held at each stage (candidate C would win the above election).

- **point system** (or **Borda count**): with five candidates, five points are awarded for each first place vote, four points are awarded for each second place vote, and so on. The candidate with the most points wins (candidate D would win the above election).

- **pairwise comparison**: the winner between each pair of candidates is determined and the candidate with the most pairwise wins is the winner of the election (candidate E would win the above election).

You can probably already see the dilemma. We have proposed five reasonable methods for finding a winner and they all give different results! The unit closes inconclusively with this question unanswered. The next unit takes up the issue of fairness in voting systems and attempts to resolve the question.

Key Words and Phrases

majority rule	popular vote	electoral vote
super majority	preference schedule	plurality
top-two runoff	sequential runoff	point system
Borda count	pairwise comparisons	

Key Concepts and Skills

- understand the concept of majority rule and apply it in two -candidate elections.
- know the difference between popular vote and electoral vote.
- use super majority rules to determine the outcome of votes.
- create and interpret preference schedules.
- apply the methods of plurality, top-two runoff, sequential runoff, point system, and pairwise comparisons to determine the outcome of elections with preference schedules.

Solutions Unit 11D

1. 1876 Presidential Election. To find a candidate's percent of the vote, divide the candidates votes by the total number of votes.

Total Popular Vote 8,319,778	Total Electoral Vote 369
Tilden 51.52%	Tilden 49.9%
Hayes 48.48%	Hayes 50.1%

Although Tilden narrowly won the popular vote, Hayes narrowly won the electoral vote and became President.

3. Other Close Presidential Elections. To find a candidate's percent of the vote, divide the candidates votes by the total number of votes.

Total Popular Vote 8,891,088	Total Electoral Vote 369
Garfield 50.04%	Garfield 58.0%
Hancock 49.96%	Hancock 42.0%

Garfield narrowly won the popular vote and easily won the electoral vote to become President.

5. To find a candidate's percent of the vote, divide the candidates votes by the total number of votes.

Total Popular Vote 79,978,736	Total Electoral Vote 537
Carter 51.05%	Carter 55.3%
Ford 48.95%	Ford 44.7%

Carter narrowly won the popular vote and comfortably won the electoral vote to become President.

7. Super Majorities.

a. The percentage of shareholders favoring the merger is $10,580/15,890 = 66.58\%$ which is just shy of the required $2/3 = 66.67\%$ vote needed.

b. A 3/4 vote of a 15-member jury is at least 12 votes (since $0.75 \times 15 = 11.25$). Thus there will be no conviction.

c. A 3/4 super majority of the states is needed to amend the Constitution. In this case, 32 of the 50 states, or 64% of the states, support the amendment, so it fails to pass.

d. A $2/3 = 66.7\%$ vote is needed in both the House and the Senate to override a veto. The override gets a $68/100 = 68\%$ vote in the Senate and a $292/435 = 67.1\%$ vote in the House. So the veto can be overturned.

9. Three Candidate Elections.

a. Best wins a plurality, but not a majority.

b. Imagine that there are 100 votes cast in the entire election. Then Able needs 18 votes to win 50% of the votes. Eighteen votes represents 18/28 = 64.3% of Crown's votes.

11. Three Candidate Elections.

a. The total number of votes cast is 465. Inviglio, who has the most votes, wins 185/465 = 39.8% of the votes, which is a plurality, but not a majority.

b. A majority of the votes is 233 votes; so Height needs 83 votes to overtake Inviglio, which is 83/130 = 63.8% of Grand's votes.

13. Ballots to Schedule. There are four different rankings of the three brands. Be sure the total votes add up to 10.

First	A	A	C	C
Second	B	C	A	B
Third	C	B	B	A
	4	2	3	1

Preference Schedules.

15.

a. A total of 66 votes were cast.

b. A received 8 first place votes, B received 20 first place votes, C received 16 first place votes, and D received 22 first place votes. Therefore, <u>D is the plurality winner</u> (but not by a majority).

c. From part (b), we see that B and D enter the runoff and the votes of A and C are redistributed. Now B receives 20 + 6 = 26 votes and D receives the remainder, or 40, of the votes. Thus <u>D is the winner of the top two runoff</u>.

d. In the sequential runoff we eliminate only the candidate with the fewest first place votes at each stage. From part (b), we see that A is eliminated first. Redistributing A's votes, D receives A's 8 first place votes; so at this point B has 20 votes, C has 16 votes, and D has 30 votes. Now C is eliminated and the election is between B and D, as in part (c). <u>The winner by sequential runoff is D</u>.

e. For the Borda Count, we score 4 points for a first place vote, 3 points for a second place vote, 2 points for a third place vote, and 1 point for a fourth place vote. The point totals are as follows:

A: $(20 \times 1) + (15 \times 3) + (10 \times 2) + (8 \times 4) + (7 \times 3) + (6 \times 3) = 156$.

B: $(20 \times 4) + (15 \times 1) + (10 \times 1) + (8 \times 1) + (7 \times 2) + (6 \times 2) = 139$.

C: $(20 \times 2) + (15 \times 2) + (10 \times 4) + (8 \times 2) + (7 \times 1) + (6 \times 4) = 157$.

D: $(20 \times 3) + (15 \times 4) + (10 \times 3) + (8 \times 3) + (7 \times 4) + (6 \times 1) = 208$.

Note that the point total is 660, as it must be. We see that <u>D is the winner by the Borda count</u>.

f. Here are the results of the 6 pairwise races:

A over B, 46 to 20.

C over A, 36 to 30.

D over A, 52 to 14.

C over B, 39 to 27.

D over B, 40 to 26.

D over C, 50 to 16.

Thus A scores 1 point, B scores 0 points, C scores 2 points and D scores 3 points. <u>D wins by the pairwise comparison method</u>.

g. As the winner by all five methods, candidate D is clearly the winner of the election.

17.

a. A total of 100 votes were cast.

b. A received 35 first place votes, B received 25 first place votes, and C received 40 first place votes. Therefore, <u>C is the plurality winner</u> (but not by a majority).

c. From part (b), we see that A and C enter the runoff and the votes of B are redistributed. Now A receives 20 votes and C receives the remainder, or 5, of the votes. A now has 55 votes and C has 45 votes. Thus <u>A is the winner of the top two runoff</u>.

d. With only three candidates, the sequential runoff method is the same as the top two runoff method.

e. For the Borda Count, we score 3 points for a first place vote, 2 points for a second place vote, and 1 point for a third place vote. The point totals are as follows:

A: $(30 \times 3) + (5 \times 3) + (20 \times 2) + (5 \times 1) + (10 \times 2) + (30 \times 1) = 200$.

B: $(30 \times 2) + (5 \times 1) + (20 \times 3) + (5 \times 3) + (10 \times 1) + (30 \times 2) = 210$.

C: $(30 \times 1) + (5 \times 2) + (20 \times 1) + (5 \times 2) + (10 \times 3) + (30 \times 3) = 190$.

Note that the point total is 600, as it must be. We see that <u>B is the winner by the Borda count</u>.

f. Here are the results of the 6 pairwise races:

B over A, 55 to 45.

A over C, 55 to 45.

B over C, 55 to 45.

Thus A scores 1 point, B scores 2 points, and C scores no points. B wins by the pairwise comparison method.

g. A wins by runoff, B wins by Borda count and pairwise comparisons, C is the plurality winner. There is not a clear winner.

19.

a. A total of 90 votes were cast.

b. A received no first place votes, B received 30 first place votes, C received no first place votes, D received 20 first place votes, and E received 40 first place votes. Therefore, E is the plurality winner (but not by a majority).

c. From part (b), we see that B and E enter the runoff and the votes of A, C, and D are redistributed. Now B picks up 20 votes, making B the winner of the top two runoff.

d. Since only three candidates received first place votes, the sequential runoff and the top two runoff methods are the same.

e. For the Borda Count, we score 5 points for a first place vote, 4 points for a second place vote, 3 points for a third place vote, 2 points for a fourth place vote, and 1 point for a fifth place vote. The point totals are as follows:

A: $(40 \times 3) + (30 \times 2) + (20 \times 4) = 260$.

B: $(40 \times 2) + (30 \times 5) + (20 \times 3) = 290$.

C: $(40 \times 1) + (30 \times 4) + (20 \times 2) = 200$.

D: $(40 \times 4) + (30 \times 1) + (20 \times 5) = 290$.

E: $(40 \times 5) + (30 \times 3) + (20 \times 1) = 310$.

Note that the point total is 1350, as it must be. We see that E is the winner by the Borda count.

f. Here are the results of the 10 pairwise races:

A over B, 60 to 30.

A over C, 60 to 30.

D over A, 60 to 30.

E over A, 70 to 20.

B over C, 90 to 0.

D over B, 60 to 30.

B over E, 50 to 40.

D over C, 60 to 30.

C over E, 50 to 40.

E over D, 70 to 20.

Thus A scores 2 points, B scores 2 points, C scores 1 point, D scores 3 points, and E scores 2 points. By the pairwise comparison method, D is the winner.

g. Candidates B and E each win two of the five methods, so the outcome is debatable.

21. Pairwise Comparison Question. With four candidates, there are $3 + 2 + 1 = 6$ pairwise races. With five candidates, there are $4 + 3 + 2 + 1 = 10$ pairwise races. In general, with n candidates, there are

$$(n - 1) + (n - 2) + \ldots + 2 + 1 = n(n - 1)/2$$

pairwise races.

23. Borda Question. Reasoning as in the previous problem, each voter will award $4 + 3 + 2 + 1 = 10$ points. If there are 25 voters, there is a total of $25 \times 10 = 250$ points. Candidates A, B, and C have a total of 120 points, so candidate D must have 130 points which is enough to win the election.

Unit 11E Theory of Voting

Overview

The previous unit closed with the observation that five reasonable methods for determining the winner of an election with three or more candidates can lead to five different winners? This unit attempts to resolve the dilemma, but as you will see, the resolution may be less than satisfying!

The analysis of voting methods involves what are known as **fairness criteria**. We will work with the following four criteria.

- **Criterion 1 (majority criterion):** If a candidate receives a majority of the first-place votes, that candidate should be the winner.
- **Criterion 2 (Condorcet criterion):** If a candidate is favored over every other candidate in pairwise races, then that candidate should be declared a winner.

- **Criterion 3 (monotonicity criterion):** Suppose that Candidate X is declared the winner of an election, and then a second election is held. If some voters rank X even higher in the second election (without changing the order of other candidates), then X should also win the second election.

- **Criterion 4 (independence of irrelevant alternatives criterion):** Suppose that Candidate X is declared the winner of an election, and then a second election is held. If voters do not change their preferences, but one or more of the losing candidates drops out, then X should also win the second election.

These criteria have been developed by researchers who study voting methods as reasonable conditions that we would expect of any fair voting system. The first criterion, stating that a candidate with a majority of the votes should win, is the most natural. The second criterion is also quite straightforward: if a candidate has more votes than each of the other candidates taken individually, then that candidate should win the election. The third criterion just says that if some voters were to change their votes in favor of the winner of the election, then that candidate should still win the election. The last criterion says that if a losing candidate drops out of the election, the original winner should remain the winner.

Most of the unit is devoted to examining the five voting methods introduced in the previous unit to see whether they satisfy the four fairness criteria. You may be surprised at the results. None of the five methods always satisfies all of the fairness criteria! This is the message of a famous result in voting theory called **Arrow's Impossibility Theorem**. It says that there is no voting method that satisfies the four fairness criteria for all preference schedules.

We next look at a method called **approval voting** that has been proposed as an alternative way to handle elections with many candidates. It departs from the familiar one-person-one-vote idea that underlies all of the voting systems discussed so far.

The unit closes with some ideas about power blocks in voting. An example of this idea is the formation of **coalitions** which is common not only in our own government, but even more so in foreign governments.

Key Words and Phrases

fairness criteria	Arrow's Impossibility Theorem	approval voting
coalition		

Key Concepts and Skills
- understand the four fairness criteria.
- apply the four fairness criteria to each of the five voting methods for a given preference schedule.
- understand the benefits and disadvantages of approval voting.
- analyze an election in terms of coalitions.

Solutions Unit 11E

3. Point System and Criterion 1. We see that candidate A has a majority of the votes (3/5 = 60%). By the point system (Borda count) A has 6 points, B has 7 points, and C has 2 points. So B wins by the point system and Criterion 1 is violated since the majority winner has lost.

7. Plurality and Criterion 2. The following preference schedule is just one possible example.

First	B	A	C	C
Second	A	B	A	B
Third	C	C	B	A
	2	4	2	3

C is the plurality winner, but A beats C and B beats C in one-on-one races.

9. Sequential Runoff and Criterion 2. By the sequential runoff method (which is the same as the top two runoff method with three candidates), candidate C is eliminated and A wins the runoff. Candidate A also wins head-to-head races against B and C, so Criterion 2 is satisfied.

11. Sequential Runoff and Criterion 2. The following preference schedule is just one possible example.

First	B	A	C
Second	A	C	A
Third	C	B	B
	3	2	4

Candidate C wins in a sequential runoff. But A beats B and C in head-to-head races, violating Criterion 2.

13. Point System and Criterion 2. The following preference schedule is just one possible example.

First	A	D	C
Second	B	B	B
Third	C	A	D
Fourth	D	C	A
	4	8	3

Using the Borda count (with 3, 2, 1, and 0 points), A receives 20 points, B receives 30 points, C receives 13 points, and D receives 27 points, making B the winner by the Borda count. However D wins head-to-head races against all other candidates. Thus Criterion 2 is violated.

15. Sequential Runoff and Criterion 3. In the sequential runoff method, candidate B is eliminated first, then candidate A, making candidate C the winner. Now suppose that the 4 voters on the third ballot (ABC) move C up and vote for the ranking CAB. Now A is eliminated first and B wins the election. Thus Criterion 3 is violated.

19. Plurality and Criterion 4. The following preference schedule is just one possible example.

First	A	B	C
Second	B	A	B
Third	C	C	A
	3	2	4

Candidate C would win by plurality. However if candidate A drops out Candidate B wins by plurality.

21. Sequential Runoff and Criterion 4. The following preference schedule is just one possible example.

First	A	B	C
Second	B	A	B
Third	C	C	A
	4	3	5

After candidate B is eliminated, candidate A wins the runoff. However if candidate C were to drop out, B would win the election. So Criterion 4 is violated.

23. Point System and Criterion 4. The following preference schedule is just one possible example.

First	A	B
Second	C	A
Third	B	C
	2	3

Using the point system (with 2, 1, and 0 points), candidate A receives 7 points, candidate B receives 6 points, and candidate C receives 2 points. This makes candidate A the winner. However if candidate C were to drop from the race, then candidate B would receive 3 points and candidate A would receive 2 points (notice that the point values become 1 and 0 with only two candidates). Thus Criterion 4 is violated.

25. Pairwise Comparison and Criterion 4. The following preference schedule is just one possible example.

First	B	C	D
Second	A	A	C
Third	D	D	B
Fourth	C	B	A
	2	2	2

By the pairwise comparison method, A would beat D, B would beat A, C would beat A and B, and D would beat B and C. Thus would lead to a tie between C and D. However, if A were to drop out of the election, then D would have two pairwise wins and C would have one pairwise win. Criterion 4 is violated because the outcome of the election is changed when A drops out.

Fairness Criteria.

27. A top two runoff would eliminate C and D. Candidate A would win the runoff, 23 votes to 18 votes. There is no majority winner, so Criterion 1 does not apply. Candidate A beats all other candidates one-on-one and is also declared winner by the top two runoff method; so Criterion 2 is satisfied. If candidate A is moved up in any of the rankings, it doesn't affect the outcome of the election, so Criterion 3 is satisfied. If either B, C, or D drops out of the race, the outcome is not changed. So in this case (but not in general), Criterion 4 is satisfied.

29. Candidate A wins by the point system with 83 points. There is no majority winner, so Criterion 1 does not apply. Candidate A beats all other candidates one-on-one and is also declared winner by the point system; so Criterion 2 is satisfied. The point system always satisfies Criterion 3. If either B, C, or D drops out of the race, the outcome is not changed. So in this case (but not in general), Criterion 4 is satisfied.

Fairness Criteria.

31. Candidate A wins by a plurality, but not by a majority. There is no majority winner, so Criterion 1 does not apply. Candidate E beats all other candidates in head-to-head races, but loses by the plurality method, so Criterion 2 is violated. The plurality method always satisfies Criterion 3. If B, C, or D were to drop out of the election, the outcome of A winning would change, so Criterion 4 is violated.

33. The candidates E, D, and B are eliminated sequentially leaving a final runoff between A and C, which candidate C wins. There is no majority winner, so Criterion 1 does not apply. Candidate E beats all other candidates in head-to-head races, but loses by the sequential runoff method, so Criterion 2 is violated. If candidate C is moved up in any of the rankings, the outcome is not affected, so Criterion 3 is satisfied. If candidate A were to drop out of the election, candidate D would still win, so Criterion 4 is violated.

35. We have seen that candidate E beats all others in pairwise races and is the winner by the pairwise comparison method. There is no majority winner, so Criterion 1 does not apply. Candidate E beats all other candidates in head-to-head races, and wins by the pairwise comparison method, so Criterion 2 is satisfied. The pairwise comparison method always satisfies Criterion 3. It can be shown that if either candidate A, B, C, or D drops out of the race, then the winner is still E; so Criterion 4 is satisfied.

37. Approval Voting.

a. Voting for only their first choices, candidate C wins by plurality with 42% of the vote.

b. By an approval vote, 28% + 29% = 57% of the voters approve of candidate A, 28% + 29% + 1% = 58% approve of candidate B, and 42% of the voters approve of candidate C. The winner by approval vote is candidate B.

12. THE POWER OF NUMBERS

Overview

This concluding chapter includes topics that have useful mathematical ideas and also represent important current issues. Of the many possible topics that might have been selected, we settled on four: the federal budget, energy, density and concentration, and logarithmic scales. Hopefully these topics will give you one last memorable look at mathematics at work in the world around us.

Unit 12A Balancing the Federal Budget

Overview

Several times in this book, particularly in problems about large numbers, we have referred to the federal budget or the federal debt. There is no question that the federal budget is a source of truly huge numbers – numbers so large that most people, including politicians, don't even have a sense of their size.

In this chapter, we give a brief survey of the essentials of the federal budget. While the details can be confusing, the general principles of the budget are straightforward. The government has **receipts** or **income** (primarily from taxes) and it is has **outlays** or **expenses** (that cover everything from defense to education to social security). In a given year, if outlays exceed receipts, then the government has a **deficit**. On the other hand, if receipts exceed expenses, then the government has a **surplus**. If the government has a deficit then it must borrow money and in the process goes into **debt**. The devastating effect of going into debt is that **interest** must be paid on the borrowed money. This interest payment itself becomes an expense category in the budget. The U.S. government typically spends 15-20% of its budget on interest on the debt. If the government has many consecutive deficit years (as has been the case in the United States), then the debt continues to grow and cannot be paid off until there are surplus years.

In this unit we clarify the ideas of deficit and debt by looking at the budget of an imaginary small business. We then consider the federal budget for the last five years and see how the major spending categories have contributed to a deficit in each year. With many deficit years, it also becomes clear how the debt has grown to its current level of about $5 trillion.

A few paragraphs are spent explaining a subtle point in reading federal budget reports. If you look closely at the federal budget summaries in this unit, you will see that the increase in the debt from one year to the next is *not* equal to the deficit for the current year, as one might expect. The explanation lies in the distinction between the **net deficit**, **net debt**, **gross deficit**, and **gross debt**, which is explained in the text.

In the last sections of the unit, we try to be somewhat practical and investigate what it will take to balance the budget (zero deficit) in a single year. A balanced budget can be achieved either by increasing receipts or decreasing outlays or both. However, not all outlays can be reduced: **mandatory** expenses must be paid and **entitlements** are very difficult to cut. This leaves only **discretionary** expenses that are eligible for cuts.

Having seen what it takes to balance the budget or have a surplus, we can ask what it will take to retire the federal debt. Some assumptions and projections are made and we predict when the federal debt will be eliminated.

The federal budget, together with its deficits, surpluses, and debts, effects everyone in many ways. Hopefully this unit will give you a basic understanding of this important issue.

Key Words and Phrases

receipts	outlays	surplus
deficit	debt	interest
net deficit	net debt	gross deficit
gross debt	mandatory expenses	entitlements
discretionary spending		

Key Concepts and Skills

- explain the terms receipts, outlays, surplus, and deficit.
- given a budget for a company or government, compute the surplus or deficit.
- understand how debt arises in a budget and propagates from year to year.
- explain the difference between net and gross deficit and debt.
- make forecasts of outlays, receipts, and deficits/surplus given rates of increase.

Solutions Unit 12A

1. The Wonderful Widgit Company Future. All figures will be rounded to the nearest dollar. Recall that figures in the table are in thousands of dollars.

a. The company must pay interest on the $773,000 debt of 1998. At a rate of 8.2% the interest payment will be $0.082 \times \$773,000 = \$63,000$.

b. The total outlays for 1999 is the sum of operating expenses, benefits, security, and interest or $600,000 + $200,000 + $250,000 + $63,000 = $1,113,000. Since receipts for 1999 were $1,050,000, the company has a deficit of $63,000. This increases the long-term debt of the company to $773,000 + $63,000 = $836,000.

c. The interest payment for 2000, on a 1999 debt of $836,000 at a rate of 8.2%, will be $0.082 \times \$836,000 = \$69,000$.

d. Under these assumptions, the total outlays for 2000 will be $600,000 + $200,000 + $0 + $69,000 = $869,000. With receipts of $1,100,000, the company has finally achieved a surplus of $1,100,000 – $869,000 = $231,000. This surplus can be used to reduce the debt to $836,000 – $231,000 = $605,000.

3. Analysis of the Federal Budget. Total outlays for 1997 were $1601 billion and the interest on the debt was $244 billion. Therefore the percentage of the total outlays spent on debt service was

$$\frac{\$244 \text{ billion}}{\$1601 \text{ billion}} = 0.152 = 15.2\%.$$

Total outlays for 1997 were $1601 billion and the defense budget was $270 billion. Therefore the percentage of the total outlays spent on defense was

$$\frac{\$270 \text{ billion}}{\$1601 \text{ billion}} = 0.169 = 16.9\%.$$

Total outlays for 1997 were $1601 billion and the cost of social security was $367 billion. Therefore the percentage of the total outays spent on social security was

$$\frac{\$367 \text{ billion}}{\$1601 \text{ billion}} = 0.229 = 22.9\%.$$

Total outlays for 1997 were $1601 billion and cost of international affairs was $15 billion. Therefore the percentage of the total outlays spent on international affairs was

$$\frac{\$15 \text{ billion}}{\$1601 \text{ billion}} = 0.0094 = 0.94\%.$$

In just three budget categories (interest, defense and social security) we can account for about 55% of the government outlays.

5. Interest on the Federal Debt. We see from Table 12.2 that in 1997 the interest payment was $244 billion, based on the 1996 debt of $5.182 trillion. This implies an interest rate of

$$\frac{\$244 \text{ billion}}{\$5.182 \text{ trillion}} = 4.71\%.$$

If the interest rate had been one percentage point higher (5.71%), the interest payment would have been

$$\$5.182 \text{ trillion} \times 5.71\% = \$296 \text{ billion}.$$

If the interest rate had been one percentage point lower (3.71%), the interest payment would have been

$$\$5.182 \text{ trillion} \times 3.71\% = \$192 \text{ billion}.$$

This problem assumes that the debt remains constant over an entire year, which isn't really the case.

7. Net Deficit and Gross Debt. The gross debt increased from $4.921 trillion to $5.182 trillion between 1995 and 1996, an increase of $261 billion. However, the net deficit was $107 billion, which is much less than the increase in the gross debt. The gross debt reflects both the net deficit (money borrowed from the public) and money borrowed from various trust funds, such as Social Security. Thus, in 1996 about

$$\$261 \text{ billion} – \$107 \text{ billion} = \$154 \text{ billion}$$

was borrowed from trust funds.

9. Budget Forecast. Because we are given a starting number and a percentage growth rate for both mandatory spending and receipts, we can calculate a future value by using the exponential growth law from (see Example 2 of the text). This equation takes the form

$$Q = Q_0 \times (1 + r)^t,$$

where Q is the value of the quantity after time t, Q_0 is the starting value of the quantity, and r is the fractional growth rate.

To forecast the mandatory and entitlement spending, we set $Q_0 = \$902$ billion for the 1996 spending, $t = 20$ years because we are forecasting 20 years later in 2016, and $r = 0.06$ for the 6% growth rate. The spending in 2016 is

$$Q = \$902 \text{ billion} \times (1 + 0.06)^{20} = \$2893 \text{ billion}.$$

For the receipts, we set $Q_0 = \$1453$ billion in 1996, $t = 20$ year, and $r = 0.04$ for the 4% growth rate. The receipts in 2016 are

$$Q = \$1453 \text{ billion} \times (1 + 0.04)^{20} = \$3184 \text{ billion}.$$

Note that, in contrast to Example 2 of the text, receipts will exceed mandatory and entitlement spending by

$$\$3184 \text{ billion} – \$2893 \text{ billion} = \$291 \text{ billion}.$$

11. Per Capita Debt. Let's take the U.S. population to be 270 million and assume that half of the population (135 million) works. The 1997 gross debt is $5.370 trillion.

Converting all these numbers to scientific notation, the per capita debt is

$$\frac{\$5.370 \times 10^{12}}{1.35 \times 10^8 \text{ people}} = \$39,778 \text{ per person.}$$

We see that if the federal debt were spread out over every worker, it would amount to over $41,000 per worker.

13. Year of Zero Debt.

a. If receipts increase at a constant annual rate of 2% and outlays are held constant every year (0% increase), the model predicts that the debt will be zero in the year 2022. These assumptions are not realistic. Outlays tend to increase every year, if for no other reason than inflation.

b. If receipts increase at a constant annual rate of 1% and outlays decrease at a rate of 3% every year, the model predicts that the debt will be zero in the year 2013. These assumptions are even less realistic than those in part (a). Real reductions in outlays are impossible to realize without significant changes to the way the government operates.

c. If the rate of increase of outlays is greater than the rate of increase of receipts, then the debt will never be eliminated. Under these conditions, spending always exceeds income and there is no way to "catch up" and pay off the debt.

Unit 12B Energy: Our Future Depends on It

Overview

This unit may seem like it belongs in a physics book. But you will see that the goal is to introduce you to the aspects of energy that affect you most immediately in your every day life; and that includes everything from reading your utility bill to planning how to eat. The mathematics in the unit is not difficult, as you will encounter many familiar ideas such as scientific notation and use of units.

The unit begins with that fundamental principles of energy. There are three basic forms of energy:

- **kinetic energy** is the energy of motion,
- **potential energy** is stored energy, and
- **radiative energy** is energy that is transmitted in light or other forms of radiation.

Of fundamental importance is the law of **conservation of energy**: energy cannot be created or destroyed; it can only change from one form to another.

One form of energy discovered only in the 20th century is the potential energy stored in mass, often called **mass-energy**. Any object with mass can in theory be turned into energy (as stated in Einstein's famous formula $E = mc^2$). This form of energy underlies nuclear power in the forms of **nuclear fission** and **nuclear fusion**.

The terms energy and power are often confused and misused. As we will see, energy is the total amount of output of a person or a system. There are many units of energy. We will use the units of **Calories, joules, and kilowatt-hours**, because they all arise in practical problems. The Calorie is used to measure the energy contained in food, the joules is the basic energy unit defined by physicists, and the kilowatt-hour is used for electrical bills.

A joule is a small amount of energy; for example, one candy bar supplies roughly one million joules of energy or about 250 Calories. The exact conversions between these energy units are:

- 1 Calorie = 4184 joules
- 1 kilowatt-hr = 3.6×10^6 joules.

Power is the rate at which you expend energy. For example, you might expend a total of 1000 Calories on a bike ride. If the bike ride takes one hour, then your power output is 1000 Calories per hour. A common unit of power is the **watt**, which is defined as

- 1 watt = 1 joule/second.

For example, a 100-watt light bulb uses 100 joules of energy per second. A related unit of power is the **kilowatt** which is 1000 watts. If a person or a device has a power output of 1 kilowatt for 1 hour, then total energy spent is 1 kilowatt-hr.

These concepts will require some thought and practice, but they comprise all of the new ideas in this unit. The remainder of the unit is devoted to practical problems involving energy and power. Table 12.4 and the related examples are also important because they allow you to compare the energy requirements of many different activities and events. For example, did you know that the energy required for an hour of walking is about the same as the energy supplied by four candy bars? This unit may be challenging, but it's also quite interesting and relevant. You can do it!

Key Words and Phrases

kinetic energy	potential energy	radiative energy
conservation of energy	mass-energy	nuclear fission
nuclear fusion	power	Calorie
joule	watt	kilowatt-hr

Key Concepts and Skills

- describe the three basic forms of energy and the law of conservation of energy.
- explain the distinction between energy and power.
- be able to use the units of energy and power and do appropriate unit conversions
- make comparisons between different uses of energy .
- solve practical problems involving energy and power.

Unit 12B Solutions

1. Exercise Power.

a. Recall that power is defined as energy divided by time. If you spend 800 Calories in one hour, the average power is 800 Calories per hour.

b. There are 4184 joules per Calorie, so 800 Calories is equivalent to

$$800 \text{ Calories} \times \frac{4184 \text{ joules}}{\text{Calorie}} = 3.3 \times 10^6 \text{ joules}.$$

Because this energy is spent over a period of 1 hour = 3600 sec, the power used in the basketball game is 3.3×10^6 joules/3600 sec = 920 joules/sec, which is 920 watts. This is enough power to keep nine 100-watt light bulbs burning.

c. In part (b), we found that the total energy used in the basketball game is 3.3×10^6 joules.

d. From Table 12.4, we see that one candy bar supplies about 10^6 joules of energy. So it would take about 3.3 candy bars to replace 3.3×10^6 joules.

3. Operating Cost of a Refrigerator.
The refrigerator uses 350 watts = 350 joules/sec. Over a period of a year, the total energy used is

$$350 \frac{\text{joules}}{\text{sec}} \times \frac{3600 \text{ sec}}{\text{hr}} \times \frac{24 \text{ hr}}{\text{day}} \times$$

$$\frac{365 \text{ days}}{\text{yr}} \times 1 \text{ yr} = 1.1 \times 10^{10} \text{ joules}.$$

There are 3.6×10^6 joules per kilowatt-hr, so 1.1×10^{10} joules is equivalent to

$$1.1 \times 10^{10} \text{ joules} \times \frac{1 \text{ kilowatt-hr}}{3.6 \times 10^6 \text{ joules}} = 3056 \text{ kilowatt-hr}.$$

At a rate of $0.06 per kilowatt-hr, the cost of running the refrigerator for a year is

$$3056 \text{ kilowatt-hr} \times \$0.06/\text{kilowatt-hr} = \$183.$$

It costs about $183 to run a refrigerator for a year.

5. Electric Bill.

a. Using the conversion from kilowatt-hrs to joules, we see that 1250 kilowatt-hrs is the same as (1250 kilowatt-hrs) × (3.6×10^6 joules/kilowatt-hrs) = 4.5×10^9 joules.

b. If one liter of oil provides 1.2×10^7 joules (Table 12.4), then the amount of oil required to supply 1250 kilowatt-hrs is (4.5×10^9 joules) × (1 liter/1.2×10^7 joules) = 375 liters. Since 1 gallon = 3.785 liters, this is the same as

$$(375 \text{ liters}) \times (1 \text{ gal}/3.785 \text{ liters}) = 99 \text{ gallons}.$$

And since a barrel of oil is 42 gallons, this amount of oil is also (99 gallons) × (1 barrel/42 gallons) = 2.4 barrels.

c. Recall that power is the rate at which energy is used or produced. The average power used is the total energy used divided by the time elapsed. Note that

1 month = 30 days × (24 hours/day) = 720 hours.

So if you use 1250 kilowatt-hrs for a period of a month, the average power use is

(1250 kilowatt-hrs)/(720 hours) = 1.7 kilowatts.

7. Fission Power.
According to Table 12.4, the total annual U.S. energy consumption is 10^{20} joules and the fission of a kilogram of uranium provides 5.6×10^{13} joules. The amount of uranium needed to supply this energy is 10^{20} joules/(5.6×10^{13} joules/kg) = 1.8×10^6 kg.

9. Energy Comparisons.
The figures in Table 12.4 are used for this problem.

a. The energy in a hurricane is on the order of 10^{17} joules while the energy released in a 1-megaton hydrogen bomb is 5×10^{15} joules. The energy in a hurricane is (10^{17} joules)/(5×10^{15} joules) = 20 times greater than the energy released in the bomb.

b. A one-hour run requires 4×10^6 joules and a candy bar provides 10^6 joules of energy. The number of candy bars needed to fuel a one-hour run is

(4×10^6 joules)/(10^6 joules/candy bar) = 4 candy bars.

c. The Sun's annual energy output is 10^{34} joules. In a second the sun releases

$$10^{34} \frac{\text{joules}}{\text{year}} \times \frac{1 \text{ yr}}{365 \text{ days}} \times \frac{1 \text{ day}}{24 \text{ hr}} \times \frac{1 \text{ hr}}{3600 \text{ sec}} = 3.2 \times 10^{26} \frac{\text{joules}}{\text{sec}}.$$

The annual U.S. energy needs are 10^{20} joules. So the total energy from the Sun in one second could supply the annual U.S. energy needs a million times over.

11. Furnace Power in BTUs.
An energy output of 1000 BTU is equivalent to

1000 BTU × 1055 joules/BTU = 1.05×10^6 joules.

A power output of 1000 BTU/hr is equivalent to

$$1000 \frac{\text{BTU}}{\text{hr}} = 1.05 \times 10^6 \frac{\text{joule}}{\text{hr}} \times \frac{1 \text{ hr}}{3600 \text{ sec}} =$$

$$292 \frac{\text{joule}}{\text{sec}} = 292 \text{ watts}.$$

(Recall that 1 watt = 1 joule/sec). The power output of the furnace is just less than 300 watts.

13. A Power Plant.
A gigawatt is 10^9 watts or 10^9 joules/sec.

a. To determine how much energy is generated in one month, we must convert months to seconds. We see that one month is

$$1 \text{ month} \times \frac{30 \text{ day}}{\text{month}} \times \frac{24 \text{ hr}}{\text{day}} \times \frac{3600 \text{ sec}}{\text{hr}} = 2.6 \times 10^6 \text{ sec.}$$

The energy generated in one month at a rate of one gigawatt is $(10^9 \text{ joule/sec}) \times (2.6 \times 10^6 \text{ sec}) = 2.6 \times 10^{15}$ watts. This number is a little more accessible if we convert it to kilowatt-hrs. We find that the energy generated in one month at a one gigawatt rate is

$$(2.6 \times 10^{15} \text{ joules}) \times (1 \text{ kilowatt-hr/}3.6 \times 10^6 \text{ joules}) = 7.2 \times 10^8 \text{ kilowatt-hr.}$$

b. If one home uses 1000 kilowatt-hr each month then a one-gigawatt power plant (such as the one in part (a) that generates 7.2×10^8 kilowatt-hr per month), can supply

$$(7.2 \times 10^8 \text{ kilowatt-hr})/(10^3 \text{ kilowatt-hr/home}) = 7.2 \times 10^5 \text{ homes.}$$

The power plant can serve over 700,000 homes.

c. Recall that 1 liter of oil supplies 1.2×10^7 joules or 3.3 kilowatt-hrs of energy. Thus, the one-gigawatt power plant that generates 7.2×10^8 kilowatt-hr per month requires $(7.2 \times 10^8 \text{ kilowatt-hr}) \times (1 \text{ liter/}3.3 \text{ kilowatt-hr}) = 2.2 \times 10^8$ liters of oil for a month. In barrels of oil this amounts to

$$2.2 \times 10^8 \text{ liter} \times \frac{1 \text{ gal}}{3.785 \text{ liter}} \times \frac{1 \text{ barrel}}{42 \text{ gal}} = 1.4 \times 10^6 \text{ barrels.}$$

The power plant requires 1.4 million barrels of oil each month or

$$(1.4 \times 10^6 \text{ barrels/month}) \times (1 \text{ month/}30 \text{ days}) = 46,667$$

barrels per day.

P 14. Nuclear Power Plants. We need to use the facts that 1 megawatt = 10^6 watts and that 1 watt = 1 joule/sec.

a. Operating at a power of 330 megawatts, the power plant generates 330×10^6 joules/sec. Converting units tells us that the number of seconds in a month is

$$1 \text{ month} \times \frac{30 \text{ day}}{\text{month}} \times \frac{24 \text{ hr}}{\text{day}} \times \frac{3600 \text{ sec}}{\text{hr}} = 2.6 \times 10^6 \text{ sec.}$$

Over a period of a month, the power plant generates $(3.3 \times 10^8 \text{ joule/sec}) \times (2.6 \times 10^6 \text{ sec}) = 8.6 \times 10^{14}$ joules. Knowing that 1 kilowatt-hr = 3.6×10^6 joules, this amount of energy is equivalent to

$$(8.6 \times 10^{14} \text{ joules}) \times (1 \text{ kilowatt-hr/}3.6 \times 10^6 \text{ joule}) = 2.4 \times 10^8 \text{ kilowatt-hrs.}$$

b. If a typical household uses 1000 kilowatt-hrs each month, then the energy from the power plant will supply $(2.4 \times 10^8 \text{ kilowatt-hrs})/(1000 \text{ kilowatt-hrs/household}) = 2.4 \times 10^5$ households or 240,000 households for a month.

c. By Table 12.4, we see that one kilogram of uranium U-235 can produce 5.6×10^{13} joules of energy by fission. This amounts to

$$5.6 \times 10^{13} \text{ joules} \times (1 \text{ kilowatt-hr/}3.6 \times 10^6 \text{ joules}) = 1.6 \times 10^7 \text{ kilowatt-hr.}$$

Therefore the 2.4×10^8 kilowatt-hrs of part (a) could be produced by

$$2.4 \times 10^8 \text{ kilowatt-hrs} \times (1 \text{ kg U-235/}1.6 \times 10^7 \text{ kilowatt-hr}) = 15 \text{ kg U-235.}$$

About 15 kilograms or 33 pounds of U-235 can produce a months worth of energy for 100,000 households.

15. Energy from Junk Mail.

a. An estimate of two pounds of junk mail per week or about 100 pounds of junk mail per year for every adult is conservative. This estimate includes home and work place. Let's assume that half of the people in the country receive junk mail (130 million people); this amounts to 1.3×10^{10} pounds of junk mail per year.

b. Using the conversion factor 2.2 pounds = 1 kg = 1000 gm, we see that the annual load of junk mail weighs $(1.3 \times 10^{10} \text{ pounds}) \times (1000 \text{ gm/}2.2 \text{ pounds}) = 6 \times 10^{12}$ gm. We are given the fact that burning a gram of paper releases 2×10^4 joules of energy. So the energy in a year's worth of junk mail is $(6 \times 10^{12} \text{ gm}) \times (2 \times 10^4 \text{ joules/gm}) = 1.2 \times 10^{17}$ joules or 3.3×10^{10} kilowatt-hours.

c. If the generation of this much energy were spread out over a year, the effective power would be (remember that power is energy per time)

$$\frac{3.3 \times 10^{10} \text{ kwatt-hr}}{1 \text{ yr}} \times \frac{1 \text{ yr}}{365 \text{ day}} \times \frac{1 \text{ day}}{24 \text{ hr}} = 3.8 \times 10^6 \text{ kilowatts.}$$

The power from paper burning is about 3.8 million kilowatts or 3.8 gigawatts. Thus paper burning could supply almost four times the power of a one-gigawatt power station.

d. The U.S. power need of 400 gigawatts is 400 gigawatts/3.8 gigawatts = 105 times the power in junk mail. So realistically (and considering pollution effects) junk mail is not a significant source of energy.

17. Wood for Energy?

a. Be sure to note the definitions of the huge units of energy that are used in this problem. Of the 180,000 terawatt-years of solar energy that reach the Earth each year, 1% of 0.06% is stored by plants. This amounts to a fraction of $0.01 \times 0.0006 = 6 \times 10^{-6}$ that is stored by plants. Thus, the total energy stored by plants is $(180,000 \text{ terawatt-years}) \times (6 \times 10^{-6}) = 1$ terawatt-year. To convert this to joules we can first convert to energy units of watt-hours (note the use of units and conversion factors):

$$1 \text{ terawatt-yr} \times \frac{10^{12} \text{ watt}}{\text{terawatt}} \times \frac{365 \text{ day}}{\text{yr}} \times \frac{24 \text{ hr}}{\text{day}} = 9 \times 10^{15} \text{ watt-hr.}$$

Note that 9×10^{15} watt-hours is 9×10^{12} kilowatt-hours. With the conversion factor 1 kilowatt-hour = 3.6×10^6 joules, we find that 1 terawatt-year is 3.2×10^{19} joules. This is the amount of solar energy that is stored in plants each year.

b. Now assume that all of the stored solar energy of part (a) (1 terawatt-year) could be made available. Over the course of a year, the average power output would be one terawatt.

c. The world power demand is 10 terawatts, whereas the power available in plants is 1 terawatt. Stored plant power could supply one-tenth (10%) of the world's power needs.

d. Even if we could somehow access the stored energy in plants, it would supply only a tenth of the world's power needs. We could generate more power by burning entire trees (representing energy stored from previous years), but plant life would be quickly depleted. Therefore, we have turned to fossil fuels that represent energy stored by plants over thousands of years.

19. Nuclear Fission Bomb.

a. Table 12.4 tells us that the fission of a kilogram of U-235 generates 5.6×10^{13} joules of energy.

b. A ton of TNT generates 5×10^9 joules. Therefore, the number of tons of TNT needed to equal the fission of one kilogram of U-235 is $(5.6 \times 10^{13}$ joules$) \times$ (1 ton/5×10^9 joules$) = 11,200$ tons.

c. Table 12.4 tells us that a 1-megaton hydrogen bomb releases 5×10^{15} joules. This energy is $(5 \times 10^{15})/(5.6 \times 10^{13}) = 89$ times more energy than that released by the fission of a one kilogram sample of U-235.

d. The 20-kiloton Hiroshima bomb is the equivalent of 20 kilotons = 20,000 tons of TNT. From part (b), a ton of TNT generates 5×10^9 joules. Therefore 20,000 tons of TNT generates $(2 \times 10^4) \times (5 \times 10^9$ joules$) = 10^{14}$ joules. The Hiroshima bomb was $10^{14}/(5.6 \times 10^{13}) = 1.8$ times more powerful than the fission bomb of part (a).

e. The implications of this calculation are clear. A one-kilogram (2.2-pound) load of U-235 could be used to make an easily-transportable terrorist bomb with over half the energy of the Hiroshima bomb.

Unit 12C Density and Concentration

Overview

The title of this unit may not seem too informative, but it does describe exactly what this unit is about. It turns out that the two related ideas of density and concentration appear in many different contexts that seem to be totally unrelated. We will talk about pollution, mining, drunkenness, and populations, all under the topic of density and concentration! This unit is a good example of how mathematics unites very different areas of study.

Density and concentration both measure the amount of some substance contained in some fixed volume; so in most general terms, density and concentration measure mass per unit volume. Usually, but not always, density is used for solid material (density of the Earth or ore or people) while concentration is used for liquids or gases (alcohol in the blood or pollutants in the air).

We first give several examples of density in practical applications. Typical units of density are pounds per cubic foot, grams per cubic centimeter, or people per square mile. Needless to say, comfort with units is important in this unit!

Units of concentration vary with the application. For example, the alcohol content of a person's blood is measured as the ratio of the volume of alcohol to the total volume of blood; it is usually expressed as a percentage. A 2% blood alcohol content means that 2% of the blood is alcohol by volume. The concentration of a pollutant in a stream is measured as the amount of material dissolved in the stream for a given volume of stream water; for example, a stream containing 5 grams of pollutant per liter of water has a concentration of 5 grams per liter. Air pollution is typically measured in parts per million (ppm) or parts per billion (ppb). If an air sample has a carbon dioxide concentration of 9 parts per million, then there are 9 molecules of carbon dioxide for every 1 million molecules of air.

Having presented these basic ideas, the unit consists entirely of examples of putting the ideas to work. With some practice, you will learn many new and important applications of mathematics.

Key Words and Phrases

Density	concentration	blood alcohol concentration
parts per million slack time	latest start time	latest finish time

Key Concepts and Skills

- understand the concepts of density and concentration, and when they are used.
- given appropriate information, compute density and concentration for the many situations described in the text.

Unit 12C Solutions

1. Calculating Densities. In all of these problems, we must use the definition of density: the mass of an object divided by the volume that it occupies. The answer should have units of mass per volume (for example, grams per cubic centimeter).

a. The density is $(0.25 \text{ kg})/(15 \text{ cm}^3) = 0.017 \text{ kg/cm}^3 = 17 \text{ g/cm}^3$.

b. The density is $(200 \text{ gm})/(7 \text{ liters}) = 28.6 \text{ gm/liter}$. Recall that a liter is 1000 cubic centimeters, so we could also write the density as 0.0286 gm/cm^3.

3. Granite and Iron. Recall that the weight of an object can be found by multiplying its density by its volume. Also the volume of an object is its weight divided by its density.

a. The volume of the granite slab is

$(100 \text{ cm}) \times (100 \text{ cm}) \times (2 \text{ cm}) = 20,000 \text{ cm}^3$.

We have converted the dimensions to centimeters because the density is given in units of gm/cm³. The weight is

$(20,000 \text{ cm}^3) \times (2.7 \text{ gm/cm}^3) = 5.4 \times 10^4 \text{ gm} = 54 \text{ kg}$.

Because 1 kg = 2.2 pounds, the weight is 119 pounds.

b. If the density of the slab were 7.9 gm/cm³ (that of iron) it would have a mass of

$(20,000 \text{ cm}^3) \times (7.9 \text{ gm/cm}^3) = 1.6 \times 10^5 \text{ gm} = 160 \text{ kg}$.

5. Population Density. Density of people is defined a little differently. Because people tend to occupy areas, their density of given by number of people per area (for example, people per square mile).

a. The population of New Jersey is 7.7×10^6 people distributed over 7419 square miles. The corresponding population density is $(7.7 \times 10^6$ people $)/(7419$ mile$^2) = 1038$ people per square mile. The population of Wyoming is 4.5×10^5 people distributed over 9.7×10^5 square miles. This results in a population density of $(4.5 \times 10^5$ people $)/(9.7 \times 10^5$ mile$^2) = 0.5$ people per square mile. New Jersey is one of the most densely populated states and Wyoming is one of the most sparsely populated states.

7. DVD Density. Density of a computer disk is measured in bytes (or megabytes) per area.

a. Recall that a gigabyte is 1 billion bytes. Therefore, a 10-gigabyte disk that has an area of 90 cm^2 has a density of $(10 \times 10^9$ bytes$)/(90$ cm$^2) = 1.1 \times 10^8$ bytes/cm^2 or 0.11 gigabytes/cm^2.

b. Recall that a megabyte is 1 million bytes. From part (a) we see that 1 square centimeter of disk holds 0.11 gigabytes or 110 megabytes of data. If a 500-page book can be stored in one megabyte, then a square centimeter of disk can hold 110 typical 500-page books. A full disk with an area of 90 cm^2 could hold approximately $90 \times 110 = 9900$ 500-page books.

9. Blood-Alcohol Concentration: Wine.

a. Example: A 155-pound person weighs about 70 kilograms. With 70 cm^3 of blood per kilogram of body weight, such a person would have 4900 cm^3 of blood. Recall that 1 liter = 1000 cm^3. So this person would have about 4.9 liters of blood.

b. There are two steps in answering this question. First, how much alcohol would result in a blood alcohol concentration of 0.1%? With 4.9 liters of blood, a 0.1% alcohol concentration requires 0.001×4.9 liters = 0.0049 liters of alcohol. Second, how much wine with a 13% concentration would you have to drink to produce 0.0049 liters of alcohol? This is a "backwards percentage problem." We ask: 13% of how many liters equals 0.0049 liters? The answer is found by dividing. It takes $(0.0049$ liters$)/(0.13) = 0.04$ liters of wine to produce this amount of alcohol. We see that 0.04 liters or about 1.4 ounces produces a blood alcohol concentration of 0.1%. This may seem like too *little* wine to bring the blood alcohol level to 0.1%. The calculation assumes that all of the wine goes directly and immediately into the blood, which is generally not the case.

c. We could repeat the calculation of part (b) with the blood alcohol limit of 0.1% replaced by 0.5%. Or we could note that the amount of wine required to reach a level of 0.5% is 5 times greater than that required to reach a level of 0.1%. Therefore, about 0.2 liters or about 7 ounces of wine are needed to reach a level of 0.5%.

d. We have seen that a 0.1% blood alcohol concentration is produced by about 0.005 liters of alcohol or 5 milliliters of alcohol. Therefore a 0.2% blood alcohol concentration results from 10 milliliters of alcohol. If drinking stopped at a 0.2% level and the alcohol were removed at a rate of 1 milliliters/hr, it would take 5 hours to reduce the blood alcohol level from 10 milliliters to 5 milliliters.

11. Lead Chloride Contamination. Given the concentration of the solution and the volume of a sample, we can find the amount of lead chloride in the sample. The weight of lead chloride is

$$0.0002 \frac{gm}{ml} \times \frac{1000 \text{ ml}}{liter} \times 1500 \text{ liters} = 300 \text{ gm}.$$

A third of a kilogram (2/3 of a pound) of lead chloride sounds like a lot! Whether it poses a health risk depends on the toxicity of the chemical and how fast it is flowing out of the mine.

13. Gaseous Pollution. Recall that the units *ppm* mean molecules of the gas per million molecules of air. A carbon monoxide concentration of 9 ppm is the same as 9000 ppb (parts per billion) since if we consider 1000 times more air molecules, we expect to find 1000 times as many carbon monoxide molecules. Therefore, the ratio of carbon monoxide to ozone molecules is

(9000 ppb of carbon monoxide)/(120 ppb of ozone) = 75.

In *any* sample of air, there are on average 75 times as many carbon monoxide molecules as ozone molecules.

15. Plutonium Release. We are told that plutonium levels were at 1.5 µg/m^3 at the plant and decrease by 0.5 µg/m^3 with each mile. One mile downwind from the plant the levels were 1.0 µg/m^3 and 2 miles downwind from the plant the levels were 0.5 µg/m^3 which is the EPA's "safe" level. This means that, within the assumptions of this model, five miles from the plant the levels are safe.

17. Stellar Corpses: White Dwarfs and Neutron Stars.

a. Using the volume formula for a sphere with a radius of 6400 kilometers the white dwarf has a volume of

$$(4/3) \times \pi \times (6400 \text{ km})^3 = 1.1 \times 10^{12} \text{ km}^3.$$

This means the density (mass divided by volume) of the white dwarf Sun will be $(2 \times 10^{30}$ kg$)/(1.1 \times 10^{12}$ km$^3) = 1.8 \times 10^{18}$ kg/km^3. To express this density in units of kg/cm^3, recall that 1 km = 10^5 cm, so 1 km$^3 = (10^5$ cm$)^3 = 10^{15}$ cm^3. The density of the Sun will be $(1.8 \times 10^{18}$ kg/km$^3) \times (1$ km$^3/10^{15}$ cm$^3) = 1.8 \times 10^3$ kg/cm^3 — a modest 1800 kg/cm^3.

b. To find mass, we multiply volume by density. Therefore, a teaspoon or 4 cm^3 of the white dwarf would have a mass of (4 cm^3) × (1800 kg/cm^3) = 7200 kg. This teaspoon of stellar matter will weigh almost eight tons!

c. With a mass 1.4 times that of the Sun, the neutron star has a mass of 1.4 × (2 × 10^{30} kg) = 2.8 × 10^{30} kg. With

a radius of 10 km, the volume of the neutron star is (4/3) × π × (10 km)^3 = 4200 km^3. The density of this star is (2.8 × 10^{30} kg)/(4200 km^3) = 6.7 × 10^{26} kg/km^3 = 6.7 × 10^{11} kg/cm^3. A cubic centimeter of this material would weight 6.7 × 10^{11} kg — ten times more than Mt. Everest.

Unit 12D Logarithmic Scales: Earthquakes, Sounds, and Acids

Overview

In this unit, our approach is to introduce logarithms in context – to show three specific applications in which it makes practical sense to use logarithms. In each of these applications, logarithms are useful because some quantity (the strength of earthquakes, the loudness of sounds, and the acidity of solutions) varies over a tremendous range – many orders of magnitude. In cases like this, logarithms always simplify the work. You may want to refer to Unit 6C where the mathematical essentials of logarithms are presented.

The first use of logarithms is to measure the strength of earthquakes. The measure of earthquake strength is called the **magnitude** and we denote it M. The amount of energy released by the earthquake (in joules) is denoted E. Clearly, the greater the energy E, the larger the magnitude M. The relationship between E and M is given by

$$\log_{10} E = 4.4 + 1.5M$$

or, equivalently,

$$E = (2.5 \times 10^4) \times 10^{1.5M}.$$

Notice that one relation involves a logarithm ($\log_{10} E$), while the second equivalent relation involves an exponential term ($10^{1.5M}$).

As shown in Table 12.5, the magnitude of real earthquakes extends from 0 to about 8. An increase by a factor of 30 in the energy of an earthquake corresponds to an increase of 1 on the magnitude scale.

The second application of logarithms is to measure the loudness of sounds. In this case, we use the **decibel** scale (denoted dB). The decibel measure of a sound is related to the intensity of the sound by

$$\text{loudness in dB} = 10 \log_{10}\left(\frac{\text{intensity of the sound}}{\text{intensity of softest audible sound}}\right)$$

or

$$\frac{\text{intensity of the sound}}{\text{intensity of softest audible sound}} = 10^{\frac{\text{loudness in dB}}{10}}.$$

Again notice that one relation involves a logarithm, while the second equivalent relation involves an exponential term. On the decibel scale, if the intensity of a sound increases by a factor of 10, the loudness in decibels increases by 1 unit. Table 12.6 gives the loudness of several typical sounds ranging in loudness from 0 dB to 140 dB.

A consideration in measuring the intensity of a sound is the distance from the source of the sound. Because sound spreads out in space as it propagates, it follows an **inverse square law**. All this means is that the intensity of a sound decreases as the *square* of the distance from the source. For example, if you increases your distance from the source by a factor of 2, the intensity *decreases* by a factor of $2^2 = 4$; if you increases your distance from the source by a factor of 10, the intensity decreases by a factor of $10^2 = 100$.

Finally, we look at the measurement of the acidity of solutions (for example, a polluted stream or a kitchen cleanser or a vinegar salad dressing). The measure that is used in this case is called the **pH**. A pH value between 1 and 6 means the solution is an acid, a pH of 7 means the solution is neutral (like pure water); and a pH between 8 and 14 means the solution is a base – the opposite of an acid. But what is it that makes a solution acid or base?

It turns out the that acidity of a solution depends on the concentration of hydrogen ions in the solution. We don't need to get into chemical technicalities, but the concentration of hydrogen ions is measured in **moles per liter** of solution, where a mole is 6×10^{23} ions. We denote the hydrogen ion concentration as $[H^+]$. With this definition, we can define the pH scale:

$$pH = -\log_{10}\left[H^+\right] \quad \text{or} \quad \left[H^+\right] = 10^{-pH},$$

As before we see both a logarithm and an exponential term appearing in this definition of pH. An increase by a factor of 10 on the hydrogen ion concentration means an increase by 1 unit on the pH scale.

Having defined the earthquake magnitude scale, the decibel scale, and the pH scale, the unit does nothing more than present examples of how these scales are used. Study these examples carefully and let them show you how logarithms work. Once you get used to logarithms, they really do simplify matters considerably.

Key Words and Phrases

earthquake magnitude	decibel scale	inverse square law
neutral solution	acidic solution	basic solution
pH scale		

Key Concepts and Skills

- understand why logarithmic scales simply work with quantities that vary over many orders of magnitude.

$$E = (2.5 \times 10^4) \times 10^{1.5 \times 5.8} \text{ joules} =$$
$$2.5 \times 10^{12.7} \text{ joules} = 1.3 \times 10^{13} \text{ joules.}$$

Solutions Unit 12D

1. Thinking in Logarithmic Scales.

a. According to Table 12.3, a magnitude 2.8 earthquake is very minor; it would have very little effect.

b. According to Table 12.6, a 160 dB is "off the scale," louder than a jet at close range. The effect would be very serious.

c. According to Table 12.7, a solution with a pH of 12 is extremely alkaline. The effect would be serious.

3. Earthquake Magnitudes.
We will need the basic relation between earthquake magnitude and energy given in the text:

$$\log_{10} E = 4.4 + 1.5M.$$

The relationship that goes "the other way" is also useful:
$$E = 10^{4.4 + 1.5M} = (2.5 \times 10^4) \times 10^{1.5M}.$$

a. The energy E of an earthquake of magnitude $M = 5$ has a logarithm given by

$$\log_{10} E = 4.4 + 1.5 \times 5 = 11.9.$$

If the logarithm base-10 of E is 11.9, the energy itself is $10^{11.9}$ joules $= 7.9 \times 10^{11}$ joules.

b. The 1985 Mexico earthquake had a magnitude of $M = 8.1$. Therefore, its energy was
$$E = (2.5 \times 10^4) \times 10^{1.5 \times 8.1} \text{ joules} =$$
$$2.5 \times 10^{16.15} \text{ joules} = 3.5 \times 10^{16} \text{ joules.}$$

c. The 1960 Morocco earthquake had a magnitude of $M = 5.8$. Therefore, its energy was

5. LA and China Earthquakes.
The 1994 Los Angeles earthquake had a magnitude of $M = 6.7$ and the 1976 China earthquake had a magnitude of $M = 7.9$. The energy of the Los Angeles earthquake was
$$E = (2.5 \times 10^4) \times 10^{1.5 \times 6.7} \text{ joules} =$$
$$2.5 \times 10^{14.05} \text{ joules} = 2.8 \times 10^{14} \text{ joules.}$$
The energy of the China earthquake was
$$E = (2.5 \times 10^4) \times 10^{1.5 \times 7.9} \text{ joules} =$$
$$2.5 \times 10^{15.85} \text{ joules} = 1.8 \times 10^{16} \text{ joules.}$$
The China earthquake was
$$(1.8 \times 10^{16} \text{ joules})/(2.8 \times 10^{14} \text{ joules}) = 64$$
times stronger than the Los Angeles earthquake. An earthquake of magnitude 7.9 would cause less damage in California than in China because of the precautions that have been taken in California to minimize earthquake damage.

7. The Decibel Scale.

a. According to Table 12.6, the sound of a siren at 30 meters is 10^{10} times louder than the softest audible sound.

b. If the intensity of a sound is 18 trillion (1.8×10^{13}) times greater than the softest audible sound, the loudness in decibels is

$$\text{loudness in dB} = 10\log_{10}\left(1.8 \times 10^{13}\right) = 133.$$

The sound could be damaging to your ears.

c. If the intensity of a sound is 1000 times less than the softest audible sound, the loudness in decibels is

$$\text{loudness in dB} = 10\log_{10}\left(10^{-3}\right) = -30.$$

The negative loudness means that this sound is too soft to be heard.

d. A 125-dB sound is

$$\frac{\text{intensity of the sound}}{\text{intensity of softest audible sound}} = 10^{\frac{125}{10}} = 3.2 \times 10^{12}$$

times more intense than the softest audible sound. Similarly, a 95-dB sound has an intensity that is

$$\frac{\text{intensity of the sound}}{\text{intensity of softest audible sound}} = 10^{\frac{95}{10}} = 3.2 \times 10^{9}$$

times more intense than the softest audible sound. Comparing the two sounds, we can say that the 125-dB sound is $(3.2 \times 10^{12})/(3.2 \times 10^{9}) = 1000$ times more intense than the 95-dB sound.

e. A sound at the pain threshold with a loudness of 90 dB is 10^9 times louder than the softest audible sound. Therefore, a sound 2 million times louder than a 90 dB sound is 2×10^{15} times louder than the softest audible sound. Using the definition of decibels, this louder sound has a loudness of

$$\text{loudness in dB} = 10\log_{10}\left(2 \times 10^{15}\right) = 153.$$

This sound is louder than a jet at 30 meters.

9. Sound and Distance.

a. According to Table 12.6, the loudness of traffic on a busy street is 80 dB measured at the edge of the street. At a distance of 100 meters from the street, 100 times farther away, the sound will be $100^2 = 10^4$ times less intense. Because each step of 10 on the decibel scale represents a factor of 10 in intensity, the traffic noise will be 80 dB − 40 dB = 40 dB.

b. The 135-dB level of the speakers is 15 dB louder or $10^{1.5} = 31.6$ times more intense than the 120-dB sound that can cause damage to the ear. Since the sound intensity decreases as the square of the distance, you should move $\sqrt{31.6} = 5.6$ times farther from the speaker or 5.6×10 meters = 56 meters away.

c. The 20-dB conversation is 100 times louder than the softest audible sound. Since the intensity decreases as the square of the distance and you are sitting 8 times further away than the conversants, the sound will be 1/64 less intense or 100/64 = 1.56 times louder than the softest audible sound. By the formula for decibels, the sound you hear will have a loudness of $10 \times \log_{10}$ (1.5) = 1.76 dB. To amplify this sound to a level of 60 dB, the dB level must increase by about 58 dB, which

means an increase in intensity by a factor of $10^{58/10} = 10^{5.8} = 6.3 \times 10^{5}$.

11. The pH Scale.

a. Remember that an *increase* of 1 unit on the pH scale means that a solution becomes 10 times more basic because the hydrogen ion concentration decreases by a factor of 10. A *decrease* of 1 unit on the pH scale means that a solution becomes 10 times more acidic because the hydrogen ion concentration increases by a factor of 10.

b. The hydrogen ion concentration of a solution with pH of 8.5 is $10^{-8.5} = 3.2 \times 10^{-9}$ moles per liter.

c. The pH is defined as

$$\text{pH} = -\log_{10} \text{ (hydrogen ion concentration)}.$$

Therefore, the pH of a solution with a hydrogen ion concentration of 0.1 moles per liter is $-\log_{10}(0.1) = -\log_{10}(10^{-1}) = 1$. This solution is an acid.

13. Toxic Dumping in Acidified Lakes.

a. With a pH of 4, the lake initially has a concentration of 10^{-4} moles of hydrogen ions per liter.

b. If the lake initially had a pH of 7, its concentration would be 10^{-7} moles of hydrogen ions per liter. The number of moles of hydrogen ions in the lake would be

$$10^8 \text{ gal} \times \frac{3.785 \text{ liters}}{\text{gal}} \times \frac{10^{-7} \text{ moles}}{\text{liter}} = 38 \text{ moles}.$$

The 100,000 gallons of pollutant with a pH of 2 (containing 10^{-2} moles per liter) contains

$$10^5 \text{ gal} \times \frac{3.785 \text{ liters}}{\text{gal}} \times 10^{-2} \frac{\text{moles}}{\text{liter}} = 3800 \text{ moles}.$$

To find the hydrogen ion concentration of the lake after the pollution, we divide the total number of moles by the total volume giving us a concentration of

$$\frac{3800 \text{ moles} + 38 \text{ moles}}{3.8 \times 10^8 \text{ liters} + 3.8 \times 10^5 \text{ liters}} = 1 \times 10^{-5} \frac{\text{moles}}{\text{liter}}.$$

The lake has a pH of 5 after the disaster.

c. We now assume that the lake already has acid water with a pH of 4 (or 10^{-4} moles per liter). The number of moles of hydrogen ions in the lake is

$$10^8 \text{ gal} \times \frac{3.785 \text{ liters}}{\text{gal}} \times 10^{-4} \frac{\text{moles}}{\text{liter}} = 3.8 \times 10^4 \text{ moles}.$$

The 100,000 gallons of pollutant with a pH of 2 (containing 10^{-2} moles per liter) contains

$$10^5 \text{ gal} \times \frac{3.785 \text{ liters}}{\text{gal}} \times 10^{-2} \frac{\text{moles}}{\text{liter}} = 3800 \text{ moles}.$$

To find the hydrogen ion concentration of the lake after the pollution, we divide the total number of moles by the total volume giving us a concentration of

$$\frac{3.8 \times 10^4 \text{ moles } + 3800 \text{ moles}}{3.8 \times 10^8 \text{ liters} + 3.8 \times 10^5 \text{ liters}} = 1.1 \times 10^{-4} \frac{\text{moles}}{\text{liter}}.$$

The lake has a pH of 3.96 after the disaster.

d. A test with a sensitivity of 0.1 on the pH scale could detect the dumping of chemicals only in the case of part (b) in which the pH drops from 7 to 5. In the case of part (c), the change in the pH is much less than 0.1.